CAMBRIDGE LIBRARY COLLECTION

Books of enduring scholarly value

Monographs of the Palaeontographical Society

The Palaeontographical Society was established in 1847, and is the oldest Society devoted to study of palaeontology worldwide. Its primary role is to promote the description and illustration of the British fossil flora and fauna, via publication of an authoritative monograph series. These monographs cover a wide range of taxonomic groups, from microfossils, trilobites and ammonites through to Coal Measure plants, mammals and reptiles, and from all ages from Cambrian to Pleistocene. They form a benchmark for understanding the past life of the British Isles and many include the original descriptions of numerous key species. The first monograph (on the Crag Mollusca) was published in March 1848 and the Society still continues this work today. Notable authors in the series include Charles Darwin (fossil barnacles) and Richard Owen (dinosaurs and other extinct reptiles). Beginning in 2014, the Cambridge Library Collection and the Society are collaborating to reissue the earlier publications, focusing on monographs completed between 1848 and 1918.

Observations on the Structure of Fossil Plants found in the Carboniferous Strata

Edward William Binney (1812–81) was a leading authority on the geology and fossil plants of the Lancashire and Cheshire coalfields. His work contributed to the early development of our understanding of the most prominent plants of the coal measure floras, in particular the habit of the trees, the relationships among organ systems, and their internal anatomical structure. This monograph, published in four parts between 1868 and 1875, is his best-known work. It focuses on the internal anatomical organisation of roots, stems and cones of horsetails and lycopods, revealing remarkable detail of the cellular structure of the tissue systems preserved in carbonate concretions. These are beautifully illustrated, and Binney rightly pays tribute to the skills of his lapidary, Mr Cuttell, and his lithographer, Mr John Nugent Fitch. The monograph is based on Binney's own collections, much of which is now housed in the Sedgwick Museum of Earth Sciences, Cambridge.

CAMBRIDGE LIBRARY COLLECTION

Books of enduring scholarly value

Monographs of the Palaeontographical Society

The Palaeontographical Society was established in 1847, and is the oldest Society devoted to study of palaeontology worldwide. Its primary role is to promote the description and illustration of the British fossil flora and fauna, via publication of an authoritative monograph series. These monographs cover a wide range of taxonomic groups, from microfossils, trilobites and ammonites through to Coal Measure plants, mammals and reptiles, and from all ages from Cambrian to Pleistocene. They form a benchmark for understanding the past life of the British Isles and many include the original descriptions of numerous key species. The first monograph (on the Crag Mollusca) was published in March 1848 and the Society still continues this work today. Notable authors in the series include Charles Darwin (fossil barnacles) and Richard Owen (dinosaurs and other extinct reptiles). Beginning in 2014, the Cambridge Library Collection and the Society are collaborating to reissue the earlier publications, focusing on monographs completed between 1848 and 1918.

Observations on the Structure of Fossil Plants found in the Carboniferous Strata

Edward William Binney (1812–81) was a leading authority on the geology and fossil plants of the Lancashire and Cheshire coalfields. His work contributed to the early development of our understanding of the most prominent plants of the coal measure floras, in particular the habit of the trees, the relationships among organ systems, and their internal anatomical structure. This monograph, published in four parts between 1868 and 1875, is his best-known work. It focuses on the internal anatomical organisation of roots, stems and cones of horsetails and lycopods, revealing remarkable detail of the cellular structure of the tissue systems preserved in carbonate concretions. These are beautifully illustrated, and Binney rightly pays tribute to the skills of his lapidary, Mr Cuttell, and his lithographer, Mr John Nugent Fitch. The monograph is based on Binney's own collections, much of which is now housed in the Sedgwick Museum of Earth Sciences, Cambridge.

Cambridge University Press has long been a pioneer in the reissuing of out-of-print titles from its own backlist, producing digital reprints of books that are still sought after by scholars and students but could not be reprinted economically using traditional technology. The Cambridge Library Collection extends this activity to a wider range of books which are still of importance to researchers and professionals, either for the source material they contain, or as landmarks in the history of their academic discipline.

Drawing from the world-renowned collections in the Cambridge University Library and other partner libraries, and guided by the advice of experts in each subject area, Cambridge University Press is using state-of-the-art scanning machines in its own Printing House to capture the content of each book selected for inclusion. The files are processed to give a consistently clear, crisp image, and the books finished to the high quality standard for which the Press is recognised around the world. The latest print-on-demand technology ensures that the books will remain available indefinitely, and that orders for single or multiple copies can quickly be supplied.

The Cambridge Library Collection brings back to life books of enduring scholarly value (including out-of-copyright works originally issued by other publishers) across a wide range of disciplines in the humanities and social sciences and in science and technology.

Observations on the Structure of Fossil Plants found in the Carboniferous Strata

Edward William Binney

CAMBRIDGE
UNIVERSITY PRESS

University Printing House, Cambridge, CB2 8BS, United Kingdom

Cambridge University Press is part of the University of Cambridge.

It furthers the University's mission by disseminating knowledge in the pursuit of
education, learning and research at the highest international levels of excellence.

www.cambridge.org
Information on this title: www.cambridge.org/9781108084352

© in this compilation Cambridge University Press 2015

This edition first published 1868–75
This digitally printed version 2015

ISBN 978-1-108-08435-2 Paperback

This book reproduces the text of the original edition. The content and language reflect
the beliefs, practices and terminology of their time, and have not been updated.

Cambridge University Press wishes to make clear that the book, unless originally published
by Cambridge, is not being republished by, in association or collaboration with,
or with the endorsement or approval of, the original publisher or its successors in title.

THE

PALÆONTOGRAPHICAL SOCIETY.

INSTITUTED MDCCCXLVII.

LONDON:

MDCCCLXVIII—MDCCCLXXV.

OBSERVATIONS ON THE STRUCTURE OF FOSSIL PLANTS FOUND IN THE CARBONIFEROUS STRATA.

ORDER OF BINDING AND DATES OF PUBLICATION.

PAGES	PLATES	ISSUED IN VOL. FOR YEAR	PUBLISHED
General Title	—	1900	December, 1900.
1—32	I—VI	1867	June, 1868.
33—62	VII—XII	1870	January, 1871.
63—96	XIII—XVIII	1871	June, 1872.
97—147	XIX—XXIV	1875	December, 1875.

OBSERVATIONS

ON THE

STRUCTURE OF FOSSIL PLANTS

FOUND IN THE

CARBONIFEROUS STRATA.

BY

E. W. BINNEY, F.R.S., F.G.S.

LONDON:

PRINTED FOR THE PALÆONTOGRAPHICAL SOCIETY.

1868—1875.

PRINTED BY ADLARD AND SON,
BARTHOLOMEW CLOSE, E.C. AND 20, HANOVER SQUARE, W.

OBSERVATIONS

ON THE

STRUCTURE OF FOSSIL PLANTS

FOUND IN THE

CARBONIFEROUS STRATA.

BY

E. W. BINNEY, F.R.S., F.G.S.

PART I.

CALAMITES AND CALAMODENDRON.

PAGES 1—32; PLATES I—VI.

LONDON:
PRINTED FOR THE PALÆONTOGRAPHICAL SOCIETY.
1868.

PRINTED BY J. E. ADLARD, BARTHOLOMEW CLOSE.

CONTENTS.

		PAGE
INTRODUCTORY REMARKS	. . . .	1
CALAMITES AND CALAMODENDRON	. . . .	3
I. BIBLIOGRAPHY	. . . .	3
II. REMARKS ON THE SPECIMENS	. . . .	11
1. GEOLOGICAL POSITION OF SPECIMENS NOS. 1—11	. . . . .	11
2. REMARKS ON SPECIMENS NOS. 1, 2, and 4—11	. . . . .	12
3. REMARKS ON SPECIMEN NO. 3	. . . .	13
4. GENERAL REMARKS ON SPECIMENS OF CALAMITES AND CALAMODENDRON	. .	15
5. REMARKS ON ASTEROPHYLLITES AND ITS FRUCTIFICATION	. . .	17
III. DESCRIPTION OF THE SPECIMENS NOS. 1—16	. . . . .	19
IV. CONCLUDING REMARKS	. . . .	30

OBSERVATIONS

ON THE

STRUCTURE OF FOSSIL PLANTS

FOUND IN THE

CARBONIFEROUS STRATA.

INTRODUCTORY REMARKS.

WHEN we consider the great number of valuable specimens from the Coal-measures of Great Britain now in our public and private collections, and see what has been done to bring them before the world, we are led to believe that our Carboniferous Fauna and Flora, but more especially the latter, have scarcely had that attention devoted to them which their importance demands. If the curators of our public Museums would describe the specimens under their charge, private collectors describe theirs, and the Council of the Palæontographical Society lend its assistance in publishing, something useful might be effected. In addition, the aid of the colliery-proprietors should be solicited for the purpose of obtaining funds to enable the Palæontographical Society to engage the best artists. When this is done we are likely to possess a literature on our Carboniferous Fossils worthy of the first coal-producing country.

Having specimens of CALAMITES in my own cabinet, collected by myself, I have been induced to make a small beginning, trusting that some other more competent parties may be induced to follow my example. Knowing the great difficulties that have to be encountered in investigating the nature of the Plants which have formed our beds of coal, my object will be chiefly to describe them, without attempting to trace their analogies with living organisms.

Probably many Members of the Palæontographical Society have specimens in greater number and more perfect preservation than those in my collection, especially as to the branches and roots of *Calamites*, the first genus of fossil plants which it is my intention to describe, but my specimens show structure in a state of perfection that has not often been met with.

On a future occasion other genera of fossil plants may be described and illustrated should an opportunity be afforded me.

My acknowledgments are due to Mr. Cuttell, lapidary, for his skill in slicing and mounting the sections of fossil wood, and to Mr. J. N. Fitch, lithographer, for the care and truthfulness with which he has executed the plates illustrating the specimens.

CALAMITES AND CALAMODENDRON.

I. BIBLIOGRAPHY.

§ 1. During many years the genus *Calamites*, so common in our Coal-measures, was generally considered to be a reed-like plant, and hence its name. Very excellent figures of the different species of this genus, with their roots and branches, are given by M. Adolphe Brongniart in his 'Histoire des Végétaux fossiles,' and by Messrs. Lindley and Hutton in the 'Fossil Flora.' All their specimens, however, gave little, if any, evidence of the internal structure of the plant. Afterwards Brongniart, in his 'Tableau des Genres de Végétaux fossiles,' after reviewing the labours of Cotta, Unger, Petzholt, and others, thought it better to divide the genus *Calamites* into *Calamitea* and *Calamodendron*, evidence having been obtained of the outer woody cylinder of the latter, which was not believed to occur in the former.

§ 2. Afterwards Mr. J. S. Dawes, who obtained much more perfect specimens than the Continental authors appear to have possessed, gave a most useful paper on the subject, which is printed in vol. vii of the 'Quarterly Journal of the Geological Society,' and he there states (p. 198) that " on lately examining a specimen of *Sigillaria reniformis*, the tissues appear so much to resemble those of the Calamite as to prove the close connection of these two genera ; in fact, all those fossils of this family with the broad outer zones of woody tissue, such as *Calamitea striata* of Cotta, will in all probability prove to be some species of small-ribbed *Sigillaria*."

About the same time Dr. Dawson gave a description of upright *Calamites* found by him near Pictou, Nova Scotia,[1] but he does not adduce any evidence as to their structure or nature.

§ 3. The two last-named authors did not appear to be aware of the publication of a paper by me " On Fossil *Calamites* found standing in an erect position in the Carboniferous Strata near Wigan, Lancashire."[2] In that communication, after describing at length the specimens exposed in the railway-cutting at Pemberton Hill, about two miles

---

[1] 'Quarterly Journal of the Geological Society,' vol. vii, p. 194, 1851.

[2] Read before the Literary and Philosophical Society of Manchester, July 6, 1847, and printed in the 'London and Edinburgh Philosophical Magazine,' ser. 3, vol. xxxi, pp. 259-266.

west of Wigan, I stated, "In the course of his examination of upright stems of *Sigillariæ* in the Coal-measures, the writer has nearly always found *Calamites* associated with them. At St. Helen's they were abundant, and their bases were found in contact with the main roots of *Sigillariæ*. One of the authors of the 'Fossil Flora,' Mr. Hutton, in describing the Burdiehouse fossils, at page 24, vol. iii, of that work, states as follows:—Amongst vegetables the characteristic fossils of this deposit are *Lepidostrobi*, *Lepidophyllites*, *Lepidodendra*, and *Filicites;* the rarity of *Calamites*, which occur but seldom and of a diminutive size, and the almost entire absence of *Stigmaria*, are very striking to those who are accustomed to view the fossil groups usually presented by the beds of the Carboniferous formation; whilst the profusion of *Lepidostrobi* and *Lepidophyllites*, of various sizes and various stages of growth, associated with the stems of *Lepidodendra*, and those of no other plant, is an additional argument for the opinion, which has always appeared highly probable, that they were the fruit, leaves, and stem of the same tribe of plants Of *Sigillaria*, a plant which in the flora of the Carboniferous group generally is of so much importance, we could not observe a trace.

"In the course of his own observations the writer has never yet been able to meet with a stem of *Sigillaria* of so small a size as six inches in diameter,[1] or a *Calamites* of so large a size as that. Doubtless there must have been young *Sigillariæ*, whether or not there were large *Calamites*. Now, what are young *Sigillariæ?* This is a question which yet remains to be answered.

"It is now admitted that little is known about the true nature of the genera *Sigillaria* and *Calamites*, except that they were not the hollow succulent stems which they were once supposed to be.

"The rootlets of *Calamites*, as previously shown, if not actually identical with, at least very much resemble, those of *Sigillaria*. In some specimens of the former genus, especially of the species *approximatus*, figured and described in pl. ccxvi, vol. iii, of the 'Fossil Flora,' and the *cruciatus*, figured in pl. xix of Brongniart's 'Histoire des Végétaux fossiles,' their rootlets were arranged in regular quincuncial order. In the largest *Calamites* that to my knowledge has been figured, namely, that called *gigas*, pl. xxvii in Brongniart's work before alluded to, the ribs and furrows begin to appear very like those of *Sigillaria*, and the joints show indistinctly. The termination of the root of a *Calamites* is exactly of the same form as the terminal point of a *Stigmaria*, both being club-shaped.

"I am not aware that up to the present time much, if anything, is known of the structure of *Calamites;* but if it should resemble that of *Sigillaria*, it may tend to prove that *Calamites* are but young *Sigillariæ*. In our observations it must not, however, be lost sight of that no central axis or pith has, to my knowledge, yet been discovered in the stem of *Calamites* like that found in *Sigillaria*. Both plants are proved to have

---

[1] Since the writing of this paper the author has seen in the Museum at Dudley a stem like that of a *Sigillaria* not more than seven inches in diameter.

had similar *habitats*, and therefore it is very probable that they might have had rootlets resembling each other without being the same plant. Still, however, as *Sigillaria* was so long considered as a separate plant from *Stigmaria*, it is unphilosophical to take no notice of the analogies of what are now considered distinct genera. Although it will not by any means be safe to affirm that *Sigillaria* and *Calamites* are the same plant, from their analogies, still it is conceived that sufficient evidence has been adduced in this paper to prove that the latter as well as the former plants have generally grown on the places where they are now found, and that the reason why one is so much more frequently found in an erect position than the other arises from the circumstance of the stem of the one being much stronger than that of the other. A deposit of mud on the branches and leaves of a slender stem of a *Calamites* might weigh it down and prostrate it, whilst the stout trunk of the *Sigillaria* would resist such action and continue erect."

As the specimens near Wigan showed the best roots of *Calamites* found standing as they grew that ever came under my observation, I herewith give in the annexed woodcut (fig. 1) a drawing (reduced one eighth the natural size) made on the spot by my friend the late M. Jobert.

FIG. 1.

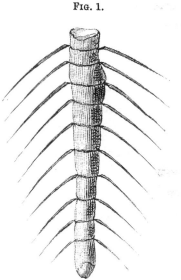

§ 4. Dr. C. von Ettingshausen, in 1855, published a most elaborate memoir on *Calamites*,[1] and showed the state of our knowledge at that time of the plant so far as its fruit, branches, and stem were concerned. This author greatly simplified the matter by classing twenty-three species of stems under *Calamites communis*, and subsequent investigations have, in great measure, if not entirely, confirmed his views; but he gives us little information on the structure of the plant, his specimens having been apparently only casts.

§ 5. Dr. Ludwig has given to the world a very lucid description, illustrated by most beautiful plates, of what he considers to be the fructification of *Calamites*.[2] His specimens are in a remarkable state of preservation, and are found associated (but not connected) with a small stem resembling a *Calamites*; this stem, however, scarcely reminds us of the plant as commonly known by that name and so plentifully found in our Coal-measures. The genus *Calamites* has, no doubt, been made to include several plants of very different structures, having external resemblance to each other in some of their characters, and Ludwig's specimen must be taken as the fructification of one of them.

[1] 'Abhandlungen der Kaiserlich-Koniglichen Geologischen Reichsanstalt, Jahrgang,' 1855. Vienna.
[2] "Calamiten-Früchte aus dem Spatheisenstein von Hattingen an der Ruhr," von Rudolph Ludwig; 'Dunker und von Meyer's Palæontographica,' vol. x, 1861 to 1863.

§ 6. Mr. J. W. Salter, F.G.S., in speaking of the Fossil Flora of the Lancashire Coal-fields, says,[1] "The great *Calamites* are common. And this genus appears to have been the most hardy, persistent, and widely diffused of the Palæozoic plants. It began with the *Lepidodendron* in the Devonian, perhaps a little later, and continued far beyond that genus into Triassic times.

"What we see of *Calamites* has been shown by several authors to be only the fluted cast of the pith of *Calamodendron*. But there is much reason to believe that the outer coating was usually very thin, and that the great succulent pith occupied nearly the whole of the stem.

"*Pinnularia* is the root of *Calamites*, as I have convinced myself by specimens during this Survey; and we have now also in Britain the fruit of this interesting genus, such as has been illustrated by Ludwig in vol. x of the 'Palæontographica.'" Mr. Salter gives a drawing of the fruit of *Calamites* from the Lower Coal-measures, Rochdale. The specimen I have not seen; but in the woodcut it exactly resembles the restored one of Ludwig.

§ 7. Professor Schimper, of Strasburg, has been one of the latest authors on the Continent who has treated of *Calamites*.[2] His specimens do not appear to afford much evidence of structure, but his remarks will give us some idea of the opinions there current as to the nature of the plant, and furnish us with the state of knowledge at the time possessed by so distinguished a botanist, who has devoted much attention to the study of both recent and fossil plants, and whose sources of obtaining information on the subject are varied and extensive.

M. Schimper's description is as follows:

"*EQUISETACEÆ*.

"CALAMITEÆ.

"CALAMITES, *Suck*.

"*Caulis cylindraceus, fistulosus, articulatus articulis clausis, sulcatis, sulcis continuatis vel in singulis articulis alternantibus, foliis in vaginam dentatam coalitis vel radiatim patentibus vel nullis (?), eorumque loco tuberculis (foliorum deciduorum pulvinulis seu foliis rudimentariis ?) minutis verticillatim dispositis.*

"Tige cylindrique ou à peu près creuse, articulée, à articles fermés par un diaphragme, sillonnée régulièrement dans le sens de la longueur; sillons se continuant directement à

[1] 'Memoirs of the Geological Survey;' "The Geology of the Country around Bolton-le-Moors, Lancashire," p. 43, 1862.

[2] "Description des Espèces de Plantes rencontrées dans le Terrain de Transition des Vosges; par Wm. Ph. Schimper." 'Mémoires de la Société des Sciences naturelles de Strasbourg,' vol. v, p. 323, 1862.

travers les articulations ou alternants ; feuilles verticillées, réunies pour former une gaîne cylindrique ou étalée, plus ou moins profondément dentée, ou nulles (?) et remplacées par des tubercules disposés en verticille (coussinets des feuilles très caduques ou feuilles rudimentaires ?).

" *Subgen.*—Asterocalamites, Schpr.

" *Vaginis profunde dentatis, divisionibus radiatim patentibus ; caulis sulcis continuis.* Gaînes profondément dentées à divisions étalées ; sillons de la tige continus.

" Calamites radiatus, *Brgt.*

"Calamites radiatus, *Brongniart. Hist. d. végét. foss.*, I, p. 122, pl. xxvi, 1, 2, 1828.

— — *Unger. Gen. et spec. plant. foss. Vind.*, p. 44, 1850.

Equisetites radiatus, *Sternb. Verst.*, II, p. 45, 1821-38.

— — *Goeppert. Foss. Flor. d. Uebergangsgeb.*, 114, 1852.

Calamites transitionis, *Goepp. Ueber d. foss. Fl. Schles.*, in Wimmer's *Fl. Schlesiens*, II, 1841; *ejusd., Fl. d. Uebergangsgeb.*, pp. 116—118, taf. 3, 4, 39, 1852.

— — *Geinitz. Verstein d. Grauwackenform. in Sachsen*, II, S. 28, taf. 18, fig. 6, u. 7, 1853; *ejusd., Kohlenform. in Haynischen-Ebersd.*, S. 30, t. 1.

Calamites cannæformis, *Schlth.* Roemer *in* Dunker und v. Meyer's *Palæontogr.*, III, 1, taf. vii, f. 4, 1850.

Bornia transitionis. *Roem. in* Dunk. und v. Meyer's *Palæontogr.* (pro specim. jun.), III, 1, taf. vii, 7, p. 45, 1850.

" *Caulis simplex plus minusve incrassatus, centim. 1-10 in diametro metiens, articulationibus cent. 1-12 a se invicem distantibus, costis exacte parallelis, subplanis, continuis, vaginis majusculis, infra medium divisis, radiatim expansis ; rhizomate ad articulationes cicatricibus singulis rotundatis a ramis caulescentibus provenientibus notato.*

"Tiges simples, offrant un diamètre de 1 à 10 cent., à articulations distantes de 1 à 12 cent., côtes exactement parallèles, presque planes, se continuant en ligne droite à travers les articulations ; gaînes assez grandes, divisées jusqu' au-dessous de la moitié de leur longueur, rayonnantes, rhizome marqué aux articulations de cicatrices isolées, provenant de l'insertion des tiges aériennes.

"Cette espèce de Calamite, la seule de genre qui ait été rencontrée dans le terrain houiller inférieur (grauwacke) des Vosges, se reconnaît facilement à ses côtes passant sans interruption à travers les articulations, de sorte que ces dernières ne se distinguent que par un sillon circulaire, accompagné des deux côtés, dans les échantillons un peu forts, d'un renflement plus ou moins évident. La grosseur des tiges rencontrées dans notre terrain varie de 0·01m. à 0·10m. et la distance de leurs articulations se montre peu constante, tant sur les petits que sur les grands individus. La gaîne qui, par son expansion horizontale et sa division profonde, rappelle un peu la verticille des Astéro-phyllites, et qui caractérise si bien notre plante, se rencontre très-rarement, parce que, à la suite de sa position naturelle, elle doit toujours rester dans la roche après la sépara-tion de la tige et dans un sens contraire à la position de cette dernière. Pour la trouver il faudrait toujours examiner la roche ambiante et la casser perpendiculairement à la tige, juste à l'endroit qui porte la contre-empreinte d'une articulation, comme cela s'est fait pour l'échantillon conservé dans le Musée d'histoire naturelle de Strasbourg, qui a été figuré pour la première fois par M. Brongniart, dans l' *Histoire des végétaux fossiles*, pl. xxvi, et que j'ai representé de nouveau à la pl. i, fig. *d* et *c*, de ce Mémoire.

"Je crois n'avoir pas besoin d'entrer dans de longs détails pour justifier la réunion du *Calamites transitionis*, Goepp., avec la *Calamites radiatus*, Brgt. Si le hasard n'avait pas fait découvrir la gaîne de ce dernier, personne n'aurait songé à le séparer du premier et encore moins à lui assigner une place dans un genre distinct. Je ne saurais non plus me ranger de l'avis du comte Sternberg et de mon ami Goeppert, en transportant notre fossile dans le genre *Equisetites*, dont il se distingue trop par la forme et la direction de la gaîne, et par la mode de disposition des sillons. Je serais plutôt porté à y voir le prototype des *Asterophyllites*, et c'est pour cela que je propose de le considérer comme type d'un sous-genre avec le nom d'*Asterocalamites*. On n'a aucun exemple qu' un végétal ou animal une fois existant ait disparu pendant un long espace de temps, pour réapparaître plus tard. Cela serait cependant le cas pour le genre *Equisetites*, qui se serait dérobé pendant toute la durée de l'époque houillère proprement dite, pour se montrer de nouveau avec le commencement de l'époque triasique et se conserver de là jusqu' à notre époque.

"Comme en Silésie et dans d'autres localités, notre *Cal. radiatus* caractérise le terrain houiller inférieur des Vosges et y constitue un des fossiles végétaux les plus répandus, surtout dans la vallée de Thann, près de Bitschwiller, d'où nous possédons de nombreux échantillons de toute dimension (Voyez pl. i)."

§ 8. Professor Goeppert has described, under the name of *Aphyllostachys Jugleriana*, the fructification of a plant allied to *Calamites*.[1] Specimens somewhat resembling those described and figured by this learned author are now in my cabinet, and were found by me in both the upper and lower parts of the Carboniferous series, namely, associated with the Spirorbis-limestone in the Manchester coal-field at Ardwick, in the lower Brooksbottom seam of coal,[2] and in the Mountain-limestone of Holywell, in North Wales. These specimens will be described and figured. The second of the specimens above mentioned (No. 12), if not of the same species as Goeppert's, is probably of the same genus.

§ 9. Herr R. Richter, in a Memoir on the Lower Carboniferous Rocks of

---

[1] Eingegangen bei der Akademie am 11 Mai, 1864. Dresden.

[2] I am indebted to the kindness of Mr. John Aitken for this specimen.

Thuringia,[1] described the structure of *Calamites transitionis* (Goeppert) as consisting of a main woody axis of cellular parenchyma, and a lesser inner axis or hollow cylinder with transverse partitions and corresponding constrictions. The first has a smooth out-side, and is formed of cono-cylindrical layers, with indistinct radial divisions, termi-nating inwards in longitudinal ridges, which give to the smaller axis a longitudinally furrowed exterior. The woody structure and a cone-like fruit of *Calamites transitionis* are figured by Richter.

§ 10. Dr. Dawson has lately published an elaborate paper ' On the Conditions of the Deposition of Coal, more especially as illustrated by the Coal-measures of Nova Scotia and New Brunswick,'[2] wherein he gives his views on *Calamodendron* and *Calamites* in the following words (p. 134) :—" *Calamodendron.*—The plants of this genus are quite distinct from *Calamites* proper. A *Calamodendron*, as usually seen, is a striated cast, with frequent cross-lines or joints ; but when the whole stem is preserved it is seen that this cast represents merely an internal pith-cylinder, surrounded by a woody cylinder composed in part of scalariform or reticulated vessels, and in part of wood-cells with one row of large pores on each side. External to the wood was a cellular bark ; and the outer surface appears to have been simply ribbed in the manner of *Sigillaria*.

" It so happens that the internal cast of the pith of *Calamodendron*, which is really of the nature of a *Sternbergia*, so closely resembles the external appearance of the true *Calamites* as to be constantly mistaken for them. Most of these pith-cylinders of *Cala-modendron* have been grouped in the species *Calamites approximatus*; but that species, as understood by some authors, appears also to include true *Calamites* (see Geinitz's ' Steinkohlenformation in Sachsen'), which however, when well preserved, can always be distinguished by the scars of the leaves or branchlets which were attached to the nodes.

" *Calamodendron* would seem, from its structure, to have been closely allied to *Sigillaria*, though, according to Unger, the tissues were differently arranged, and the woody cylinder must have been much thicker in proportion.

" The tissues of *Calamodendron* are by no means infrequent in the coal, and the casts of the pith are common in the sandstones; but its foliage and fruit are unknown. (Fig. 31, Plate VII, *a* to *c*.)

" *Calamites.*—Nine species of true *Calamites* have been recognised in Nova Scotia, of which seven occur at the Joggins ; the most abundant being *C. Suckovii* and *C. Cistii*. The *Calamites* grew in dense brakes on sandy and muddy flats. They were unques-tionably allied to *Equisetaceæ*, and produced at their nodes either verticillate simple linear

[1] 'Zeitschrift der deutschen geologischen Gesellschaft,' Jahrgang 1864, p. 155. I am indebted to Professor Rupert Jones for calling my attention to this Memoir as these pages were passing through the press.

[2] Read December 20th, 1865. ' Quarterly Journal of the Geological Society,' vol. xxii, pp. 95—169, 1866.

leaves, as in *C. Cistii*, or verticillate branchlets with pinnate or verticillate leaflets, as in *C. Suckovii* and *C. nodosus*. The *Calamites* do not seem to have contributed much to the growth of coal, though their remains are not infrequent in it. The soils in which they most frequently grew were apparently too wet and liable to inundation and silting up to be favorable to coal-accumulation. I have elsewhere shown ('Quart. Journ. Geol. Soc.,' vol. x, p. 34) that some of the species of *Calamites* gave off numerous adventitious roots from the lower parts of their stems, and also multiplied by budding at their bases" (p. 135).

Again, at pp. 140, 141, in treating of discigerous wood-cells, Dr. Dawson says: "These are the true bordered pores characteristic of *Sigillaria*, *Calamodendron*, and *Dadoxylon*. In the two former genera the discs or pores are large and irregularly arranged, either in one row or several rows. In the latter case they are sometimes regularly alternate and contiguous. In the genus *Dadoxylon* they are of smaller size, and always regularly contiguous in two or more rows, so as to present a hexagonal areolation. Discigerous structures of *Sigillaria* and *Calamodendron* are very abundant in the coal, and numerous examples were figured in my former paper. I have indicated by the name *reticulated tissue* certain cells or vessels which may either be reticulated scalariform vessels, or an imperfect form of discigerous tissue. I believe them to belong to *Stigmaria* or *Calamodendron*. (Figs. 57 and 68, Pl. XII.)"

§ 11. Professor Dr. C. von Ettingshausen[1] has lately described and figured some interesting specimens of *Calamites* showing their branches and leaves, but not affording much evidence of their internal structure.

§ 12. When the present memoir was nearly all written, Mr. Carruthers, after seeing some of my specimens, gave a restored figure of *Calamites* in a paper published in December, 1866,[2] in which he concludes, "It is not easy to find anything analogous to *Calamites* among recent plants; nevertheless its structure does not differ so essentially from the vascular cryptogams as to cause any uncertainty as to its position. The histological character of its wood, the absence of medullary rays, and the nature of its fruit, clearly establish that it was a true cryptogam; and while it differed in the arrangement of the parts of its stem, in its foliar appendages, and in its organs of fructification, from *Lepidodendron*, yet it is evident that these were both near allies, and both more highly organized than any of their living representatives."

[1] 'Die fossile Flora des Nährisch-Schlesischen Dachschiefers; Denkscriften der Kaiserlichen Akademie der Wissenschaften,' Wien, 1866.

[2] "On the Structure and Affinities of Lepidodendron and Calamites," by W. Carruthers, F.L.S., Botanical Department, British Museum; 'Seeman's Botanical Magazine' for December, 1866. In the 'Popular Science Review' for July, 1867, p. 295, &c., Mr. Carruthers has further described and figured the Calamite and its fruit.

II. Remarks on the Specimens.

§ 1. *Geological Position of the Specimens, Nos. 1 to 11.*

The Specimens described in this Memoir are sixteen in number; and all those exhibiting structure (with the exception of No. 3, which came from South Owram) were found by me in the lower division of the Lancashire Coal-measures, imbedded in calcareous nodules occurring in seams of coal.

Specimens No. 1, 2, and 4 to 11 are from the same locality as the *Trigonocarpon* described by Dr. J. D. Hooker, F.R.S., and myself in a memoir "On the Structure of certain Limestone-nodules enclosed in seams of Bituminous Coal, with a description of some Trigonocarpons contained therein";[1] and the other specimen, No. 3, is from the same seam of coal in the Lower Coal-measures as that in which the specimens described in a paper entitled "On some Fossil Plants, showing Structure, from the Lower Coal-measures of Lancashire"[2] were met with, but from a different locality, namely, from the "Halifax Hard Seam" at South Owram, marked ** in the section given below. Specimens of *Calamodendron* are also sparingly met with in the "Bullion Mines" near Burnley, marked **,—and in the "Upper Foot Mine," near Oldham, marked *** in the following vertical sections of Coal-measures.

The position of the seams of coal in which the fossil woods were found in the Carboniferous series is shown by the following sections of the Lower Coal-measures.

In Lancashire.	Yds.	Ft.	In.	*In Yorkshire.*	Yds.	Ft.	In.
Arley or Royley Seam ...	1	1	0	Beeston or Silkstone Seam	2	0	0
Strata ...	69	0	0	Strata ...	77	0	0
Seam ...	0	0	3	Royds or Black Seam ...	0	2	10
Strata ...	57	0	0	Strata ...	38	0	0
Seam ...	0	0	6	Better Bed Seam	0	1	4
Strata ...	45	0	0	Strata ...	5i	0	0
Upper Flagstone (Upholland) ...	50	0	0	Upper Flagstone (Elland)	40	0	0
Strata ...	20	0	0	Strata ...	40	0	0
Seam (90 yards) ...	0	0	5	Seam (90 yards)	0	0	6

[1] 'Philosophical Transactions,' 1855, p. 149.

[2] 'Quarterly Journal of the Geological Society of London' for May, 1862, vol. xviii, p. 106.

In Lancashire.

	Yds.	Ft.	In.
Strata	20	0	0
Seam (40 yards)	0	1	6
Strata	64	0	0
*** Upper Foot Seam (Dog Hill)	0	1	2
Strata	15	0	0
** Gannister Seam	1	0	0
Strata	13	0	0
Lower Foot Seam (Quarlton)	0	2	0
Strata	17	0	0
Bassy Seam (New Mills)	0	2	6
Strata	40	0	0
Seam	0	0	10
Strata	10	0	0
Sand or Featheredge Seam	0	2	0
"Rough Rock" of Lancashire, "Upper Millstone" of the Geological Survey	20	0	0
Strata (Rochdale or Lower Flags)	120	0	0
*Seam	0	0	6
Strata	2	0	0
Seam	0	0	10
Strata	14	0	0
Seam	0	1	3

"Upper Millstone" of Lancashire.

In Yorkshire.

	Yds.	Ft.	In.
Strata	56	0	0
Seam (40 yards)	0	1	0
Strata	39	0	0
** Halifax Hard Seam	0	2	3
Strata	14	0	0
Middle Seam	0	0	11
Strata	24	0	0
Soft Seam	0	1	6
Strata	56	0	0
Sand Seam	0	0	4
"Upper Millstone" of Phillips, Halifax	36	0	0
Strata (Lower Flagstone)	72	0	0
Little Seam	0	0	3

In the Lancashire Coal-field all the seams of coal from the "Forty Yards" downwards have at places afforded *Aviculopecten* and other marine shells in the roofs of black shale; and these latter strata generally contain calcareous nodules. The nodules in the seams of coal commonly known by the name of "Bullions" have chiefly been found in the beds marked *, **, and *** in Lancashire; whilst in Yorkshire they have as yet been only observed in the "Halifax Hard Seam," marked **.

§ 2. *Remarks on Specimens Nos.* 1, 2, *and* 4 *to* 11.

The specimens showing structure intended to be described in this Memoir are (with the exception of No. 3) from the thin seam of coal marked * in the vertical section of the Lower Coal-measures of Lancashire previously given; and are from the same "mine" or bed from which the specimens described by Dr. Hooker and myself were obtained. They were found associated with *Halonia, Sigillaria, Lepidodendron, Stigmaria, Trigonocarpon, Lycopodites, Lepidostrobus, Medullosa,* and other genera of plants not yet determined. The foregoing are mentioned in the order of their relative abundance.

A portion of one of the specimens of fossil wood on analysis[1] gave

Carbonate of lime	...	...	...	...	76·66
Carbonate of magnesia	...	...	...	...	12·87
Sesquioxide of iron	...	...	...	...	4·95
Sulphate of iron ...	...	...	...	...	0·73
Carbonaceous matter	...	...	...	...	4·95

The stratum lying immediately above the seam of coal in which the specimens occurred generally termed the "roof," was composed of black shale, containing large calcareous nodules, and for a distance of about two feet six inches upwards was one entire mass of fossil shells of the genera *Goniatites, Orthoceras, Aviculopecten,* and *Posidonia.*

The beds in the vicinity of the coal occurred in the following order, namely,

	Yds.	Ft.	In.
1. Black shale, with nodules containing fossil shells	0	2	6
2. Upper seam of coal, enclosing the nodules full of fossil wood	0	0	6
3. Fire-clay floor, full of *Stigmaria*...	0	2	0
4. Clay and rock	2	0	0
5. Lower seam of coal	0	0	10
6. Fire-clay, full of *Stigmaria.*			

The fossil wood occurred in spherical, lenticular, and elongate and flattened nodules, varying from an inch to a foot in diameter; the round and globular specimens being in general small, whilst the flatter nodules were nearly always of a large size. No fossil shells were met with in the nodules found in the coal itself ("2"), although, as previously stated, they were very abundant in the nodules found in the roof ("1") of the seam, which there rarely contained any remains of Plants. The large nodules of ten to twelve inches in diameter, when they occurred, swelled out the seam of coal both above and below, as in the annexed woodcut, fig. 2.

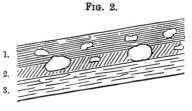

FIG. 2.

§ 3. *Remarks on No. 3 Specimen.*

The third specimen intended to be described in this Memoir is from a small seam of coal, about two feet in thickness, in the Lower Coal-measures, and marked ** in the vertical section at p. 12; and is from the same seam that the specimen of *Sigillaria vascularis* described by me in the paper published in the 'Quarterly Journal of the Geo-

[1] For this analysis I am indebted to the kindness of Mr. Hermann.

logical Society,' previously quoted, came from, although from a different locality. This specimen occurred in the " Halifax Hard Seam" or " Gannister Coal," at South Owram, near Halifax. It was associated with *Sigillaria*, *Stigmaria*, *Lepidodendron*, *Halonia*, *Diploxylon*, *Lepidostrobus*, *Trigonocarpon*, and other fossil Plants not well determined. The above is about the order of relative abundance in which these plants occurred.

A portion of one of the nodules gave on analysis[1]—

Sulphates of potash and soda	1·620
Carbonate of lime	45·610
Carbonate of magnesia	26·910
Bisulphide of iron	11·650
Oxides of iron	13·578
Silica...	0·230
Moisture	0·402

The stratum found lying immediately above the seam of coal in which the nodules occurred was composed of black shale, containing large calcareous concretions, and for about eighteen inches was one entire mass of fossil shells of the genera *Aviculopecten*, *Goniatites*, *Orthoceras*, and *Posidonia*.

The beds occurred in the following (descending) order :

	Ft.	In.
1. Black shale, full of fossil shells and containing calcareous concretions ...	1	6
2. Halifax Hard Seam, with the nodules containing the fossil plants	2	0
3. Floor of fireclay and gannister, full of *Stigmaria ficoides*.		

The fossil wood is found in nodules dispersed throughout the coal; some being spherical, and others elongated and flattened ovals, varying in size from the bulk of a common pea to eight and ten inches in diameter. In some portions of the seam of coal the nodules are so numerous as to render it utterly useless; and they are found to occur over a space of several acres, and then for the most part to disappear, and again to occur as numerous as ever. For the distance of twenty-five to thirty miles the nodules are found in this seam of coal in more or less abundance, but always containing nearly the same plants. Fossil shells are rarely met with in the nodules found in the coal; but they occur abundantly in the large calcareous concretions found in the roof of the " mines," and are there associated with *Dadoxylon* (containing *Sternbergia* piths) and *Lepidostrobus*. So far as my experience extends, the occurrence of nodules in the coal is always associated with that of fossil shells in the roof, and therefore may probably be owing to the presence of mineral matter held in solution in water and precipitated upon, or aggregated around, certain centres in the mass of vegetable matter now forming coal before the bituminization of such vegetables

---

[1] For this analysis I am indebted to the kindness of Dr. R. Angus Smith, F.R.S., who had it done in his laboratory by Mr. Browning.

took place.[1] No doubt such nodules contain a fair sample of the plants of which the seams of coal in which they are found were formed; and their calcification was most probably due to the abundance of shells afterwards accumulated in the soft mud and then decomposed, and now forming the shale overlying the coal.

At present little is known of the process by which animal and vegetable bodies are decomposed, and the particles of which they were formed removed and exactly replaced by mineral matter. All observers have been struck with the wonderful perfection of the process by which the most microscopic parts of minute vessels and cells have been preserved in form; but no author could satisfactorily account for it until the wonderful discoveries in *Dialysis* by Professor Graham, F.R.S., H.M. Master of the Mint, showed us how crystalloids, such as carbonate of lime, could percolate through animal and vegetable membranes. It is probably by the laws of *Dialysis* that we shall be enabled to find out the process of the calcification of the specimens described in this Memoir.

§ 4. *General Remarks on Specimens of* CALAMITES *and* CALAMODENDRON.

For a long time I have devoted considerable attention to the genus *Calamites*, and have collected a tolerably good suite of specimens which show structure. There is no difficulty in obtaining any quantity of fossil wood, showing the wedge-shaped bundles of pseudo-vascular[2] structure, which, springing in radiating series from certain circular and oval orifices, next the central axis, are parted by wedge-shaped masses of very coarse cellular tissue, increasing in the opposite direction to those first mentioned. It is well known that the casts of the central axis, showing by their ribs and furrows the former position of the wedge-shaped bundles, are found in most of our Coal-measures; but when we look for specimens affording outside characters of the woody cylinder, and examples of the pith or central axis showing structure, we find great difficulty in obtaining them. Most collectors select large specimens, and in these, although we may sometimes be so fortunate as to meet with evidence of the outside of the plant better than a mere carbonaceous film, generally we have not much chance of getting any indication of the pith or central axis. The wedges of pseudo-vascular tissue, proceeding from their radiating orifices, are usually found compressed close together, and the space formerly occupied by the pith or central

---

[1] For a very excellent account of the petrification of wood, see Dr. Goeppert's 'Die Gattungen der fossilen Pflanzen,' published at Bonn in 1841, a work of great research, and which does not appear to have been much known in England, judging from the few references made to it. This learned author long anticipated me in the discovery of the structure of the rootlets of *Stigmaria*, although I was quite unaware of it when I wrote my paper on the same subject published in the 'Quarterly Journal of the Geological Society,' vol. xv (1859), p. 79.

[2] The term "pseudo-vascular" is taken from Mr. Dawes' paper on *Calamites* published in vol. vii of the 'Quarterly Journal of the Geological Society.' It may be a question whether "vascular" should not be employed.

axis has been assumed to be composed of lax cellular tissue, like *Sigillaria* and *Stigmaria* were supposed to have been. Mr. Dawes, as well as Petzholt, had noticed the transverse divisions in the pith or central axis of *Calamites* at the nodes, and both authors considered the pith to be composed of cellular tissue, with a few vascular bundles in it.

Searchers after fossil plants showing structure must be aware, if they have any great experience, that large specimens have their tissue generally very much distorted and disarranged, and seldom afford any evidence of the central axis or pith, except a mere line, or a cast of mineral matter, in an amorphous condition. My endeavours have been to find very small specimens, for these appear to have undergone less alteration from pressure than larger ones. They vary in size from $\frac{6}{100}$ of an inch in diameter to three inches; and it is only in the very smallest specimens that we find every part of the plant preserved. Transverse sections of small specimens are met with which show only a radiating cylinder of pseudo-vascular tissue of two tubes in breadth, whilst some of the larger ones exhibit a radiating cylinder of upwards of one hundred tubes in breadth.

In the small specimens we have not much chance of seeing the external characters of the stem, and for this we have to resort to larger specimens.

In order to reconstruct the whole of the Plant, it is necessary to build up its parts from different individuals of various sizes. The outside of a small *Calamites* is ribbed and furrowed, and shows nodes or joints; whilst a larger specimen, in its decorticated state, is nearly smooth, or slightly marked with fine longitudinal striæ. The cell-walls of the tubes composing the pseudo-vascular cylinder of the former are thin, and the oval openings not so well defined; whilst in the latter, the cell-walls are stronger, and the elongated openings in them are of a more distinct form. In both large and small specimens the central axis or pith is divided at the joints by horizontal diaphragms. Both have a thick carbonaceous bark, and their pseudo-vascular systems are wedge-shaped, springing from orifices[1] or openings around the central axis.

In this Monograph no attempt will be made to distinguish the genus *Calamodendron* from the old genus of *Calamites;* but, as all my chief specimens show structure, they will be named *Calamodendron*. It was formerly supposed that the larger specimens belonged to the latter genus, and the smaller ones were classed with the former: but my specimens, both large and small, afford only evidence of one kind of structure; and I am, therefore, induced to class both provisionally under one genus in this Memoir.

Other observers may bring forward fresh grounds, from structure, to show that *Calamites* is a distinct plant from the *Calamodendron* herein described. My specimens I have named *Calamodendron commune*.

---

[1] As I know of no similar arrangement in living plants, I have used the words "orifices" or "openings" in the descriptions, in preference to the terms *nuclei* or *areolæ*.

§ 5. *Remarks on* ASTEROPHYLLITES *and its* FRUCTIFICATION.

For many years *Asterophyllites* has been known as the leaves of *Calamites*, and numerous specimens have been figured and described by Lindley and Hutton, Brongniart, Ettingshausen, Presl, and others. From the joints of the stem of the plant, at each of the oval spaces around it, proceeded a number of branches, radiating in every direction. These, at their joints, sent out smaller branches, which, at intervals, from lesser joints, again furnished whorls of leaves, with a length of an inch to an inch and a half. The leaves were of the shape of a *Lepidophyllum*, but scarcely so long or so broad; and were marked in the middle by a strong midrib. In my cabinet are some specimens of *Asterophyllites* with leaves longer than have been usually met with; and one of these will be figured, although not from the same locality as the specimens which show structure came from. Unfortunately no well-recognised specimens of the external forms of leaves have yet been met with in the calcareous nodules which afford the fossil stems showing structure.

The fructification of *Asterophyllites* is well shown in some specimens found by me in the red shales of the Upper Coal-measures of Ardwick, near Manchester. A specimen will be figured and described, which throws considerable light on the nature of the plant.

An example of another plant, resembling the *Volkmannia sessilis* of Presl, and allied to *Asterophyllites*, and found in the Mountain-limestone of North Wales, will be given.

By the kindness of my friend, Mr. John Aitken, of Bacup, I am enabled to give the figure of the fructification of another allied plant, found near the Lower Brooksbottom Coal (the lowest in the section previously given, p. 12) by Mr. Henry Stephenson, at Ewood Bridge. This, if not the same as the *Aphyllostachys Jugleriana* of Goeppert, is very nearly allied to it.

The three last-mentioned specimens are all about the same size so far as their cones are concerned; and these agree pretty well with the small cones found in great abundance near the stems of *Calamodendron*, except in being about twice the bulk of the latter; but all four are very short when compared with the long cones of *Flemingites* described and figured by Mr. Carruthers.[1] Most probably this last-named genus had whorls of leaves, like those of *Asterophyllites*, proceeding from a jointed stem, very similar to, but not identical with, that of *Bechera grandis*, figured and described by Lindley and Hutton.

The fructification of *Calamites* has long been supposed to be that known as *Volkmannia*. Some years since, when examining my specimens of *Calamodendron*, showing

[1] "On an undescribed Cone from the Carboniferous beds of Airdrie," 'Geological Magazine,' vol. ii, p. 433, October, 1865.

3

structures, I noticed numerous small cones lying detached around them in the stone. These I immediately recognised as resembling *Volkmannia*, and showed them to Dr. J. D. Hooker, F.R.S., who also thought that they bore great resemblance to that genus of fossil plants. On further examination, from the structure of the central axis of the cone being the same as that of the stem of *Calamodendron*, it was evident to me that the cone was the fructification of that plant. This was before Dr. Ludwig's paper came under my notice.

My specimens of the organs of fructification will be described at length, as they show structure in all their parts, and exhibit the spores in the sporangia, which Dr. Ludwig's do not, so far at least as described by that author.

The cone is one third of an inch long in my best specimens, although it may have exceeded that length. In one case there are eight sporangium-receptacles, placed one immediately above the other. These receptacles are of a crown-shape, formed by the scales which proceed from the central axis of the cone, at first at right angles, and then, when they reach the outside, taking a vertical direction, somewhat like the scales of *Lepidostrobus*, figured and described by Dr. Hooker[1]; but they are arranged one above another throughout the whole series, and not in a spiral direction. The sporangia are of an irregular egg-shape, slightly elongated, and are arranged in fours, symmetrically, around a thorn-like process or spindle coming from the axis; but I have not been able in my specimens to see them distinctly enveloped in a bladder-shaped bag so plainly as described by Ludwig: but there is clear evidence of such a covering. In each receptacle there are six of these series of four, arranged radially with regard to the central axis; so there are twenty-four sporangia altogether in every receptacle. Each sporangium has a covering composed of a single row of parallel cells, which generally shows evidence of some disturbance, so that the original form of the sporangium is not often well displayed. This is filled with numerous round spore-like bodies, some of them having apparently a tri-radiate appearance, and looking as if they had divided into three sporules. These are not unlike similar spores seen in *Lepidostrobus Browni*: but are more transparent, not so dark in colour, and of smaller size.

The attachment of the bladder-shaped bag, containing the four sporangia, to the spindle is not well seen; but the connection of the latter to the central axis of the cone is clearly shown, and is exactly the same as that described by Dr. Ludwig in his specimen.

The outside of the central axis is composed of tubes of hexagonal and pentagonal forms, having all their sides marked by transverse openings of an elongated oval shape, similar to what are observed on the pseudo-vascular bundles of tubes of *Calamodendron*; but blank spaces show that some portions of the axis have been of a more perishable nature.

---

[1] 'Mem. Geol. Surv. Great Britain,' vol. ii, part 1, p. 449, plate 7, fig. 8.

The form of the sporangium, when it is of large size, and had attained a state of maturity, is generally of an irregular egg-shape, as before stated; but, in some small specimens, it has a cordate or pear-shape form in tangential sections, and is filled full of dark-coloured spores; and its external cover, composed of a single row of cells, is thicker and more substantial. This envelope appears to have expanded and become thinner as it grew older; and, at length, when the spores were ripe, it probably burst and dispersed them. Specimens of sporangia are to be found in all stages of their growth; but the cones, although generally showing most beautiful structure, have been very much disarranged, and it is extremely difficult to get true sections of them, especially in a longitudinal direction, parallel to their central axis. A great number of specimens had to be examined before the quadrate arrangement of the sporangia, as first shown by Ludwig, could be made out, as well as the occurrence of six processes or spindles in the stead of that author's five, and the twenty-four sporangia against his twenty.

III. Description of the Specimens.

§ 1. The Specimens (*Calamodendron commune*) Nos. 1 and 2.

(No. 1, Plate I, fig. 1; Plate II, figs. 1—6. No. 2, Plate I, fig. 2.)

Specimen No. 1 (Plate I, fig. 1) is about eight inches in length, and ten inches in circumference. Although no doubt originally cylindrical, it is now of an irregular pear-shape from the pressure to which it was probably subjected in the process of mineralization. The outside of the fossil, which is in a decorticated state, is marked by fine longitudinal striæ; but it does not show the ribs and furrows so commonly found on what is generally termed the outside of *Calamites*, nor are there any joints or nodes apparent on it. There is some evidence of the outer bark in a thin coating of bright coal, the specimen having shelled out of the matrix in which it was imbedded, and left its bark in the stone. But in another specimen (No. 2; Plate I, fig. 2), from the same place as No. 1 came, we have an example of a *Calamodendron* which has been split through the middle, and which there shows, on the outside of the central axis, the usual ribs and furrows so long considered as belonging to the outside of the plant; and in an upper portion of the specimen, not shown in the Plate, is a joint.

On looking at a transverse section of No. 1 (Plate II, fig. 1), the wedge-shaped bundles of pseudo-vascular structure are seen squeezed together, and divided by other wedge-like masses of coarse cellular tissue, of a lighter colour; but there is no trace left of the central axis (or pith) of the plant. This, as has been previously stated, is nearly always the case with large specimens, so far as my experience goes. In the specimen now under consideration, owing to the position of its fracture, we cannot obtain any evidence of the cast of the outside of the central axis; but if the specimen were fractured like No. 2, it would most probably in all respects be similar to that figured in Plate I, fig. 2, as to the usual ribs, furrows, joints, and nodes so commonly found on ordinary *Calamites*.

The wedge-shaped bundles of pseudo-vascular tissue originate from a small circular orifice or opening, sometimes simple, as in the specimen now under consideration (Plate II, fig. 2), but in other instances, apparently divided into several parts, as shown in the annexed woodcut[1] (fig. 3); they are composed of quadrangular tubes, arranged in radiating

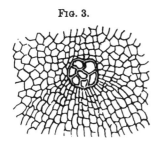

FIG. 3.

series, and increasing in size as they approach the circumference. These tubes, but hexagonal in section, also extend in an imperfect radiating series some distance from the "orifice" towards the central axis, and increase in size as they approach that point. They are divided by oblique dissepiments, and their walls are much thicker than those found in the inner woody cylinder of *Sigillaria;* and, instead of being covered with fine transverse striæ, both simple and anastomosing, as in that genus of plants, they are marked by oval openings in the sides, horizontal in the longitudinal section, Plate II, fig. 3, and termed by Dr. Dawson "reticulated tissue."[2] These openings are also shown, in a tangential section, in Plate II, fig. 4, taken near the commencement of the wedges of pseudo-vascular tissue. The diameters of the tubes increase gradually from their origin at the orifice, or opening, to their termination at the outside of the cylinder, where they are largest. The number of these wedge-shaped bundles in this specimen (No. 1) is seventy-three, alternating with an equal number of wedge-shaped masses of coarse and lax quadrangular tissue, having their broadest side next to the central axis, and diminishing in size as they extend towards the circumference; the size of the cells thus decreasing in the opposite direction to that of the decrease of the tubes of the pseudo-vascular cylinder, namely, from the central axis to the outside. These can be seen with the naked eye to about ⅟th of the distance from the central axis to the outside, where they cannot be recognised without the aid of a microscope; but tangential sections of this part of the pseudo-vascular cylinder show that it is traversed by bundles of tissue, oval in section (Plate II, fig. 6) not much unlike in shape to those seen in *Sigillaria vascularis*, but with

---

[1] This cut is from a drawing made by Mr. Bone under the direction of Dr. J. D. Hooker, who after carefully examining these openings, I believe, came to the conclusion that they were passes for a peculiar kind of tissue which has unfortunately been destroyed, rather than the mere cavities which we now see in the specimens.

[2] The openings in the walls of the tubes have a resemblance in shape to Dr. Dawson's left-hand piece in fig. 67, pl. 12 ('Quart. Journ. Geol. Soc.,' vol. xxii); but the walls of the tubes in my specimens are much thicker than those described by Dr. Dawson (op. cit., p. 140, 169). This structure, no doubt, is familiar to all who have carefully examined under the microscope the charcoal in coal.

the marked difference of being composed of cells and not of barred vessels, as is the case in that plant.

The origin of the pseudo-vascular wedges is from the oval orifices, and they seem to have no apparent connection, so far as my observation goes, with the central axis or pith. In small specimens the number of these orifices had been noticed as far as six ; whilst in the large specimen now under description, as previously stated, they amount to seventy-five in number. Specimens of these wedge-shaped masses, similar in structure and external character to those of No. 1, can be seen of different sizes from some so small as to consist of only three tubes in breadth (from the orifice to the outside) up to larger ones, with more than one hundred tubes.

The specimen shows what at first sight might be mistaken for annular rings, or depositions of successive growth ; but when carefully examining it under the microscope, we do not find in it sufficient evidence to establish with certainty any appearance of the cessation of growth, like that shown by the annular rings of an exogenous plant of the present day. Other specimens in my cabinet give more evidence of successive growths ; but still in my opinion scarcely sufficient to establish the former existence of distinct annular rings, showing the stoppage of growth, like that which now takes place in our hard-wooded exogenous trees ; and these appearances in the specimen may have been caused, at the time of the mineralisation of the specimen, by successive deposits of mineral matter. However they may have been produced, these rings appear to me to be nearly similar in *Calamodendron* to those usually seen in transverse sections of the external woody cylinder of *Sigillaria*, as well as those in the outsides of *Dadoxylon*.

The tangential section Plate II, fig. 5, is taken near the commencement of the pseudo-vascular bundles, and shows them of small size and divided by broad spaces of coarse cellular tissue. That of Plate II, fig. 6, is taken nearer to the circumference, and shows the pseudo-vascular bundles separated by others of coarse cellular tissue, elongated oval in section. Neither of these sections, however, exhibits the oval-shaped bundles of vessels proceeding from the joints, and communicating with the branches. These will be shown in other specimens, hereinafter described.

§ 2. The Specimen (*Calamodendron commune*) No. 3.

(Plate III, figs. 1—6.)

This is of small size when compared with Nos. 1 and 2 last described, and is exhibited in Plate III, fig. 1, displaying only one side and the top of the specimen. It is one inch in length, and three-tenths of an inch in diameter across its major axis. The outer bark has been converted into a film of bright coal, which adheres to the stony matrix, and thus leaves the outside of the stem in a decorticated state. This stem, unlike

Specimen No. 1, is marked with the usual longitudinal ribs and furrows, interrupted by joints, so commonly met with on the exteriors of ordinary *Calamites*. This difference in the outside appearance of large and small specimens is probably due to the extremities of the pseudo-vascular bundles forming the ribs, and the lax tissue the furrows, in the young specimens; whilst the older individuals, having an exterior composed chiefly of pseudo-vascular tissue without such marked divisions of cellular tissue, do not exhibit such decided ribs and furrows. However different in appearance the outsides of the specimens Nos. 1 and 3 are, we shall presently see that their structure is the same in nearly all respects.

Fig. 2 represents a transverse section of the stem, magnified ten diameters. It is oval in form, and shows twenty-two wedge-shaped masses of pseudo-vascular structure, radiating from twenty-two different orifices, placed outside the central axis at regular distances, and parted by wedge-shaped masses of lax tissue, increasing and diminishing in opposite directions to what obtains in the pseudo-vascular tissue, as previously noticed in the larger specimen No. 1 (p. 20). The tubes composing the latter tissue gradually increase in size as they extend from the orifices towards the circumference; whilst in the lax tissue the cells diminish as they are traced from the outside of the central axis to the exterior.

The central axis has been destroyed, with the exception of two oval-shaped portions of tissue, which, however, are too imperfectly preserved to afford us much evidence of their original structure. Information as to this we shall receive from another and smaller specimen hereinafter described.

Fig. 3 represents a longitudinal section of the pseudo-vascular part of the stem (magnified seven diameters), showing it to have been composed of quadrangular tubes, having their sides marked by oval openings, placed horizontally, like those previously described in No. 1 (p. 20). These tubes, when they approach the joints swell out; but after they have passed those parts they assume their usual size. At the joints a diaphragm of coarse cellular tissue appears to divide the stem horizontally, not much unlike what is found in *Dadoxylon*; but, of course, less frequently, and at greater distances.

Fig. 4 shows a tangential section of the stem, and affords evidence of the structure of three of the bundles (oval in section) of vessels which proceeded from the joints to the branches. These bundles, in their characters, both as to shape and structure, are like those seen traversing the internal woody cylinder of *Sigillaria vascularis*, from the inside to the leaves. This section also shows other masses, of a lighter colour than the body of the specimen, and a more elongated oval shape, composed of three and four cells in breadth. These appear to be extensions of the wedge-shaped masses of lax tissue which divide the bundles of pseudo-vascular tissue near the orifices. One of these is shown magnified forty-five diameters in fig. 6, and will afford a good idea of their general appearance.

Fig. 5 exhibits a longitudinal section of the same stem, magnified fifty diameters. With this magnifying power the walls of the tubes give evidence of oval openings, placed horizontally; but they are scarcely so well defined as those in the large specimen No. 1; and the walls appear to be slighter (even after allowing for the smaller size of the specimen), than those in the specimen first described (p. 20).

The two sections last mentioned, more especially the longitudinal one, are obscured in parts by patches of coaly matter; and, as we should expect from the appearance of the transverse section of the stem, we find that a portion of the central axis has been destroyed, and no trace of its original structure left. The coaly matter is, no doubt, owing to the line of section being taken close to the bark, which is generally found converted into bright coal.

§ 3. The Specimen (*Calamodendron commune*) No. 4.
(Plate III, fig. 7.)

Plate III, fig. 7, shows a very small specimen of *Calamodendron*, magnified thirty-four diameters, with the central axis (or pith) in a complete state of preservation. The diameter at the broadest is only $\frac{6}{100}$th of an inch. The stem, no doubt, was originally cylindrical, like those of all other specimens of this genus of plants; but it has assumed an oval form from pressure when it was in a soft state. The central axis, or pith, is composed of large pentagonal utricles (some filled with a black colouring matter), arranged without order, except that the largest in size is found near the centre. The woody cylinder surrounding the central axis consists of nine wedge-shaped masses of pseudo-vascular structure, radiating from as many orifices, and similar in all respects to that of the two larger specimens, Nos. 1 and 3, previously described (pp. 20 and 22). Indeed it appears to be an individual of the same species, in a younger stage of growth.

The form of the utricles composing the central axis somewhat resembles that found in small specimens of *Sigillaria vascularis*. In all my longitudinal sections of this small stem, I have not been able to satisfy myself that they are marked on their sides by fine horizontal striæ; but still they appear to be different from ordinary cellular tissue, and have more the form of utricles than of cells. This is a very material point to clear up, and probably the examination of other specimens may enable us to elucidate it more satisfactorily.

§ 4. The Specimen (Cone of *Calamodendron commune*) No. 5.
(Plate IV, fig. 1.)

The fructification of *Calamites* was long ago supposed by Ettingshausen and others to be the same as *Volkmannia*; and the great number of small cones, evidently nearly allied to the latter genus, found lying around my specimens, pointed out the probability of such being the case, when they first came before me; but it was only when the structure of their central axis was examined and found to be of the same character as the stem of *Calamodendron* that the connection of the one with the other was clearly established.

Plate IV, fig. 1, represents a transverse section of Specimen No. 5, one of the cones, $\frac{1}{10}$th of an inch in diameter, magnified forty-five times. It shows the central axis of the column, composed of hexagonal and pentagonal tubes, which have been somewhat displaced from their original position. Around the part last described is a space where the structure is not shown; and then comes a six-sided girdle of rather larger tubes, giving rise to bundles of pseudo-vascular tubes (enveloped in cellular tissue), which constitute the thick portion of the disc-like scale, which divides the cone into receptacles, or cells, containing the sporangia, and from which the leaves proceed and go upwards. These sporangia (marked *j j*) are of an irregular oval shape, having the broader ends near the periphery and the narrower next the central axis. They were arranged in series of fours, around a stout spike, as described by Ludwig. In this transverse section, of course, only two are shown, and they appear to have been enveloped in a bladder-shaped bag, traces of which are seen in the dark curved lines (marked *kk*) bounding the four best preserved sporangia, shown in the lowest part of the figure, and which, when in a perfect state, would in a transverse section have presented a cordate form.

In this specimen distinct evidence of six of these bags, each containing twelve sporangia, is shown instead of the five containing ten sporangia in Ludwig's figure. The outer coating of the sporangium is composed of a single row of cells, and the sporangium itself is full of round bodies, like microscopic spores, some of which have an appearance of a triradiate ridge on their outsides, but the majority appear plain, as represented in the figure. The section seems to have been made across the cone midway betwixt the base of the scale forming the cell-partition, and the spike or spine supporting the sporangia, and therefore affords little evidence of the structure of either of those parts of the cone. For this we must resort to the longitudinal sections contained in Plate V, which will afford us the requisite information.

§ 5. The Specimen (Cone of *Calamodendron commune*) No. 6.
(Plate 4, fig. 2.)

Plate IV, fig. 2, represents a transverse section of No. 6, ($\frac{1}{19}$th of an inch in diameter, magnified 54 times), another cone, similar to that last described. This has the walls of its sporangia considerably disarranged, and only a few spore-like bodies are scattered about the section. The central axis is in a fair state of preservation, and is composed of a mass of hexagonal and pentagonal tubes, smaller in size than those shown in fig. 1, but like them in other respects. A space without structure then intervenes between the axis and an irregular zone of larger tubes, from which zone project six angular arms, that appear to me to be transverse sections of six of the processes which support the sporangia, and which it will be well hereafter to term "sporangium-bearers." These seem to have been composed of tubes, having something of a pseudo-vascular character; and the lowest one in the figure appears to join with the outside covering of the bladder or bag (*k k*) containing the sporangia; but of this we cannot be certain, as the parts of the specimen are much disarranged.

These two transverse sections (Pl. IV, figs. 1 and 2), which are the best preserved amongst many in my collection, afford us evidence that the central axis or column of the cone was composed of hexagonal and pentagonal tubes, surrounded by a substance which has left no evidence of its structure, and outside this is the zone of hexagonal coarser tubes from which spring six heart-shaped bags or bladders containing the upper twelve of twenty-four sporangia, similar to those described by Ludwig, except that their number is two more than were seen in his specimen.

§ 6. DESCRIPTION OF SPECIMENS (CONES OF *Calamodendron commune*) Nos. 7—11.
Plate V, figs. 1—5.

Plate V, figs. 1, 2, 3, 4, 5, 5*a*, and 5*b* represent the specimens, Nos. 7, 8, 9, 10, and 11, and are longitudinal sections of small cones similar to those (Nos. 5 and 6) of which transverse sections are given in Plate IV, and described at pages 23 and 24.

Fig. 1 (No. 7) represents a specimen one third of an inch in length, and magnified thirteen and a half diameters. It shows seven receptacles, or cells for sporangia; and more may have existed in the original specimen, as the extremities of the cone are probably wanting in the section. The bases or pedicles of the scales belonging to the second and third receptacles (from the top) are seen to proceed on each side from the central axis, and appear to consist of tubes or utricles, whilst the other divisions or floors, four in number, which are cut outside the central axis, are concavo-convex, and appear to be composed of a thick band of cells, whence spring the stout, fleshy, outside scales or leaves, which take a vertical direction, and enclose each receptacle until the bottom of the next scale or leaf is reached. In each receptacle is seen evidence of sporangium-bearers, one opposite the other, springing from the central axis. In the uppermost chamber is a tangential section of one; in the 2nd, a similar section of two; in the 3rd, a longitudinal section of two, which are seen to proceed from the central axis; the 4th, a tangential section of two; the same in the 5th. In the 6th is a tangential section showing four sporangia grouped round a sporangium-bearer; and this affords us distinct evidence of the quadrate arrangement of the sporangia around the "bearer," as first noticed by Ludwig. In the 7th and last receptacle, shown only on the outside of the central axis, no evidence of a longitudinal section of a sporangium-bearer is seen, but a part of one and larger portions of the terminal parts of four other scales or leaves are shown.

Most of the sporangia in this cone are disarranged, or their sections are such as not to give us a good idea of their original form; but they are all full of sporelike bodies.

Fig. 2 (No. 8) is a cone rather less than one third of an inch in length, magnified fourteen diameters. It gives evidence of eight receptacles (all holding sporangia, more or less disarranged and containing a few spores), eight sets of scales or leaves, and sporangium-bearers, both in tangential and longitudinal sections, as follow :—in each of the first four receptacles, going downwards from the top, are two tangential sections: in the 5th, two longitudinal sections and one tangential: in the 6th, one longitudinal and one tangential section: in the 7th are two longitudinal sections, showing portions of the sporangium-bearers, connected with the central axis of the cone: in the 8th, there are no remains of the sporangium-bearers; but a portion of the base of a scale and the whole of the terminal part of that organ are visible.

In this figure, although the section only shows a portion of the column, the form of the scales and leaves forming the receptacles is well shown in the lower part. This specimen in all respects so exactly resembles that of fig. 1, that it requires no further description.

Fig. 3 (No. 9), magnified nineteen diameters, is a tangential section of a single receptacle, and shows four sporangia, of a cordate form, two of them being full of dark-coloured sporelike bodies. The outsides of these sporangia appear to be thicker and darker in colour than the generality of the specimens. Probably these differences may arise from the sporangia being in a younger stage of growth. The section is outside the central axis, and near to the ends of the sporangia. These dark sporelike bodies are frequently met with, in a detached state, in the nodules; but they are not often found in their position in the receptacle as seen in this case. The sporelike bodies also appear to be larger in size than those commonly met with in ordinary sporangia.

Fig. 4 (No. 10), magnified twenty-four diameters, represents a longitudinal section of little more than a single receptacle of a cone, but it clearly shows the marked difference of the scales, both in form and structure, to the sporangium-bearers. The thick fleshy bases or pedicles of the scales appear to have formed a disc- or cup-shaped division between the receptacles, and to have been composed of coarse cellular tissue on the outside, enveloping a bundle of pseudo-vascular tissue, which was prolonged into the apex of the scale or leaf. The sporangium-bearers are broad at their point of connection with the central axis, but they soon taper off, and form a spindle-shaped process like a thorn or spine, chiefly composed of pseudo-vascular structure. Some of the sporangia contain few sporelike bodies, and some are entirely empty.

Fig. 5 (No. 11), magnified eighteen diameters, represents the most perfect longitudinal section of a cone that has yet come under my observation. In it we find the structure of the central axis to be of the same character as that of the pseudo-vascular bundles in *Calamodendron* (see p. 20 and 23), both being composed of tubes that have their walls perforated with oval openings. A portion of this structure, magnified 130 diameters, is shown in fig. 5*b*, and will be at once recognised by all who have investigated the microscopical characters of coal, as frequently occurring in the charcoal or " mother-coal," and which, as far as I know, has never been clearly traced to any particular fossil plant. This specimen also affords evidence of the thick scales, or divisions of the receptacles, with their leaflike ends, and of the sporangium-bearers, all in their natural positions, connected with the central axis, and not separated and disarranged, as most generally met with. A highly enlarged sporangium (magnified forty-five diameters) full of sporelike bodies, taken from this specimen is given in fig. 5*a*.

The structure of the central axis of the cone, as seen in fig. 5, and as previously shown in specimens Nos. 5 and 6, Pl. IV, figs. 1 and 2 (p. 23, &c.), appears to have had its middle composed of tubes having a pseudo-vascular structure, surrounded by a zone of something which has not been preserved, and now shown by a blank space in the specimen. Next comes a zone of larger-sized tubes, of a hexagonal form, from which spring the receptacle divisions (or scales and leaves), and the sporangium-bearers.

§ 7. DESCRIPTION OF SPECIMENS Nos. 12, 13, 14, 15, and 16.
[Plate VI.]

In the strata near where the nodules containing the fossil wood showing structure occurred no specimens of *Asterophyllites* or of the fructification of that genus of plants were found except those previously described; but in my cabinet are three specimens, from the Carboniferous strata, of the fructification of plants most probably allied to *Calamodendron*, which are worth describing. In all their characters, but more especially as regards size and shape, they bear great resemblance to the specimens of fructification before mentioned. Unfortunately none of them afford any evidence of internal structure; we get only an outside view of the sporangia, and see nothing of their internal parts or contents, as we see in my specimens from the coal-seams.

Plate VI, fig. 1 (No. 12), represents a stout stem, having traces of ribs and furrows, and seven joints, at which knots appear. From these last-named parts, on each side of the stem, are seen to proceed seven cones, each about half an inch in length, springing outwards in a nearly horizontal direction in the specimen. These cones do not expose any trace of a central axis; but are composed of crown-shaped masses, most probably of sporangia, contained in receptacles, arranged around an axis. Eight or nine of these can be seen in one cone. Unfortunately the specimen being in soft shale, no evidence can be obtained of its internal structure, so as to ascertain if the sporangia contained any spores. If it is not the same as Dr. Goeppert's *Aphyllostachys Jugleriana*,[1] it is very closely allied to that plant.

---

[1] 'Ueber Aphyllostachys, eine neue fossile Pflanzengattung aus der Gruppe der Calamarien, so wie

As previously stated, I am indebted to my friend Mr. John Aitken for the specimen which I believe was found by Mr. H. Stephenson, near the lowest Brooksbottom Coal, at Ewood Bridge, Lancashire, about fourteen yards below the position of the greater number of the specimens described in this memoir, and near the seam of coal marked * in the vertical section of the Lancashire Coal-field previously given (p. 12). The only point in which this specimen appears to differ from Ludwig's (see above, p. 5), is that it only possesses eight to nine receptacles or cells, against his fifteen to sixteen.

Another specimen of the fructification of a plant evidently allied to *Asterophyllites* and *Calamodendron* is given in Plate VI, fig. 2 (No. 13), magnified half as large again as the original. This consists of a stout stem, finely ribbed and furrowed, and affording evidence of four sets of fruit-cones, springing upwards at a high angle from four joints of the stem. Although only four cones are seen at each joint, more may be underneath, covered up in the matrix. In two of those sets which are more perfect than the rest we observe traces of four cones, and in the other only two. Each cone has a central axis or column, from which spring ten scales on a side, forming receptacles or sporangium cases, similar to those described in Pl. VI, fig. 1 (p. 27), except that there we only see eight on each side. In this specimen we do not find the terminal point of the cone, so that there is no positive evidence of the number of scales it originally possessed.

In size and characters, especially as to its stem, this specimen bears considerable resemblance to the *Volkmannia sessilis*, of Presl, figured by Dr. Goeppert,[1] as well as to the fructification of *Calamodendron;* and although probably the evidence may not be sufficient clearly to connect either this or the last-described specimen specifically with the *Calamodendron commune* figured by me, still they must be considered as nearly allied to it, as well as to Ludwig's specimen (see above, p. 5), and they are valuable in showing the connection of the cones with the stems on which they grew.

In the Upper Coal-measures of Ardwick, near Manchester, above the highest seam of coal there met with, is an abundance of *Calamites* and *Asterophyllites*, especially the species *A. longifolia*, of Lindley and Hutton; and connected with the latter plant, and lying around both, are numerous fruit-cones, which, although showing no structure (being embedded in a liver-coloured shale), give a clear idea of their external form, and the mode by which they were connected with the stem and leaves of *Asterophyllites*.

Pl. VI, fig. 3 (No. 14), represents a specimen of *Asterophyllites longifolia* (from Ardwick) in my cabinet, magnified half as large again as its natural size. Springing from the stem, at each of the joints, twelve to fourteen verticillate leaves are shown in the part of the specimen exposed, and probably as many more may be concealed on the other side

über das Verhältniss der fossilen Flora zu Darwin's Transmutations-Theorie;' von Dr. H. R. Goeppert, Dresden, 1864.

[1] See Goeppert on 'Aphyllostachys,' *antè*.

in the matrix. Some of these leaves measure nearly an inch and a half in length, and are marked with a keel. Altogether, in form and character (except being a little less in size), those leaves bear great resemblance to *Lepidophyllum*.

In Pl. VI, fig. 4 (No. 15), is another specimen from Ardwick (magnified twice its natural size), having a stem of about two inches in length, and not so stout as the two last described, but more deeply furrowed and more sharply ribbed. At each of the joints of the stem (two of which are visible) are seen four fruit-cones, accompanied by as many leaves of *Asterophyllites longifolia*, springing outwards; and probably as many more may be concealed in the matrix underneath.

The cones consist of a central axis, on either side of which, exactly opposite each other, are seen seven or eight pairs of cordate bodies, each having a division in its middle. This organ somewhat resembles the bladder-shaped bag or envelope, containing sporangia, described by Ludwig (see above, p. 24). They are bounded by a scale or disc, coming at first from the axis nearly at right angles, but afterwards running almost parallel to it, and forming the receptacle or sporangium-cases.

In three of these cones is the termination of the scales; and there are six cones seen, probably as many more lying underneath in the shale: thus it is probable that each receptacle had six bags, containing four sporangia each; or altogether twenty-four, as in the specimens showing structure previously described (p. 24). The thorn or spindle, around which the sporangia are fixed, is not well shown unless we take the dark line of division of the sporangium-case to be it. The whole of the specimen so far as the cone is concerned, in its external characters, resembles my other specimens of the fructification of *Calamodendron commune* rather than Ludwig's specimens; and it is found, we must remember, not only with the leaves of *Asterophyllites*, connected with the stem on which it grew, but surrounded by an abundance of detached leaves and stems of that plant.

Pl. VI, fig. 4 *a* (No. 16), represents the apex of a cone, from Ardwick, magnified three diameters; and shows six leaves or scales on the side of the specimen which is exposed to view.

IV. Concluding Remarks.

After the description of the specimens of *Calamodendron* in this Monograph, it will not be out of place to notice the points in which this genus of plants differs from *Sigillaria*, especially from the plant which I have described as *Sigillaria vascularis*.[1]

The form of the roots of *Sigillaria* varies from that of *Calamodendron*. It is true that the termination of *Stigmaria* is club-shaped, like that of *Calamodendron*, and the rootlets are arranged in quincuncial order, as was shown in the paper of mine previously alluded to (see above, p. 15) ; but in the latter plant we find no trace of the regular bifurcation of the main and lesser roots ; and the roots appear to have been small in size, with regard to the stem and branches of the plant, when compared with similar parts in *Sigillaria*. From all the evidence which has been obtained by me, *Calamodendron* must have been a plant of small size when compared with *Sigillaria*.

The largest *Calamodendron* that has come under my notice is one from the Middle Coal-measures, and was found by the late Mr. John Atkinson, F.G.S., in the neighbourhood of Chesterfield ; it is now in my possession. The cast of the central axis of this specimen is five and a half inches in diameter ; and, taking the proportions of smaller specimens, the woody cylinder, exclusive of the bark, would probably be about one foot in diameter ; a small size, when compared with some stems of *Sigillaria*, which have been found to measure seven feet in diameter at the base.

The terminal branches of *Calamodendron* were also of small size, some of them not being more than $\frac{6}{100}$ths of an inch in diameter ; whereas the smallest specimens of *Sigillaria vascularis*, which have come under my notice have been about half an inch across ; and the branches, like the roots before mentioned, in *Calamodendron* have none of the dichotomous characters so distinctly shown in *Sigillaria vascularis*.

The organs of fructification did not reach more than one third of an inch in length, and were of a diminutive size even when compared with the small bulk of the plant. We cannot well compare them with similar parts of *Sigillaria*, as at present we are unable to speak with absolute certainty as to the fructification of that plant, which was most probably a cone much larger than that of *Calamodendron*. Indeed the small size of the organs of fructification in *Calamodendron* is one of the most singular characters of the plant.

There is a difference also observed in the bundles of vessels passing from the centre to the circumference, dividing the wedge-shaped masses of pseudo-vascular tissue, and which M. Adolphe Brongniart and some other authors considered to be of the nature of

---

[1] 'Quart. Journ. Geol. Soc.,' vol. xviii, p. 106, pl. IV and V.

medullary rays, but which Mr. Carruthers, with apparently very good reasons, regards as not having that character in *S. vascularis*. At the joints, where the branches spring from the stem, the bundles of vessels there found much resemble those of *S. vascularis;* but the tissue dividing the pseudo-vascular bundles in *Calamodendron* is very different; and, when seen in the small openings towards the outer part of the woody cylinder, has more the appearance of a medullary bundle (being formed of cellular tissue) than those found in *Sigillaria vascularis*, which are barred on all their sides, like the tubes forming the woody cylinder of that plant.

Although some years since it occurred to me, as well as to others, that small *Calamodendron* might possibly have been young *Sigillaria*, judging only from the rootlets and some other portions of the plants, still we have seen that in the external characters of the two genera there is no evidence to establish their identity. When also we come to compare the internal structure of the two plants, one differs from the other so much as to dispose of any outward resemblance. In *Sigillaria*, whether we take *Diploxylon cycadoideum* or *Sigillaria vascularis*, there are two woody cylinders, formed of radiating tissues, while in *Calamodendron* there is only one such woody cylinder. The tubes composing the internal radiating cylinders in the two first-named plants have their walls covered with fine horizontal lines or striæ, either free or anastomising; while in the last-named plant the walls of the tubes are much thicker, and are pierced by oval openings, having their major axes at right angles to the direction of the tubes. In addition, the central axis in *Sigillaria* has no horizontal diaphragms dividing it into separate portions, as is the case in *Calamodendron*, although the casts of the two central axes are each striated longitudinally.

From the examination of the specimens described in this Monograph, *Sigillaria* and *Calamodendron* must be considered as two distinct plants, although they doubtless grew in similar positions, and in their habitats accompanied each other, in greater or less abundance, during the whole of the time in which the Carboniferous strata were in the course of formation, as is evident from the remains of both plants being so frequently found associated together in the " mother-coal" of this and other countries.

In the upper seam of coal in the section of the strata previously given (p. 12), and there marked with a single asterisk, *Calamodendron* is by far the most common plant, and *Sigillaria* (in the form of *Diploxylon cycadoideum*) is but rarely found; while in the seams marked with two and three asterisks, higher up in the series, *Sigillaria vascularis* is by far the commonest form, and *Calamodendron* is rarely met with; *Calamodendron* being the smaller plant, forming the chief part of the small seam of coal, and *Sigillaria* the larger plant, forming the greater proportion of the thicker seam of coal. Something similar occurs in the Upper Coal-measures of Ardwick, whence the specimens Nos. 14 and 15 came. *Calamites*, then, is one of the most common plants met with, and it is found associated with *Asterophyllites, Lepidodendron, Lepidostrobus,* and *Lepidophyllum,* but with no traces of large-ribbed and furrowed *Sigillaria*, and only rare specimens of *Sigillaria elegans.* In this locality no seam of coal is found associated with the fossil plants; and a small bed, a few inches in thickness, occurs about fifty feet below them. My observations on this part of the subject have been made chiefly whilst searching for specimens affording evidence of structure, and not over any great thickness of the strata, and therefore further attention should be devoted to it in order to determine whether or no this condition occurs throughout the Coal-measures where *Sigillaria* and *Calamites* are found associated. Many years since, when examining the thick seams of coal which are found in the middle division of the Lancashire Coal-field, I noticed the occurrence of large-ribbed and furrowed *Sigillariæ*, and I stated in a paper on the Origin of Coal, read before the British Association in 1843, and printed in the Report of the Association's first Meeting at Manchester, that where such large specimens occurred not only were the seams of coal thick, but they were open-burning coals, leaving a white ash. Many years' observation on this subject has confirmed my first impression. Of course it is not contended that ribbed and furrowed *Sigillariæ* are not to be met with from the lowest to the highest Carboniferous strata and their roots (*Stigmariæ*) found in coal-floors throughout, but it is merely intended to state that during the formation of the thick seams of the Middle Coal-measures greater crops of these trees prevailed, and produced more mineral ingredients to form the white ash, as well as the thicker seams of coal.

NOTE.—The Author thinks it necessary to state that the Specimens described in this Memoir were discovered twelve years ago, and some of them immediately "mounted" and sent to Dr. Hooker, who had consented to join in publishing a description of them. With the specimens was written, November 29th, 1854, "You will be delighted to find such beautiful sections of *Volkmannia* in the large slide I now send. "They are well worth the trouble and expense of the slide. These, doubtless, belong to *Calamites*." Soon afterwards the Author had the opportunity of pointing out under the microscope the spore-like bodies in the sporangia to Dr. Hooker.

At that time the whole of the structure of *Calamodendron* had been made out, with the exception of the centre of the stem and the connection of *Volkmannia* with it. The valuable paper of Dr. Ludwig on the Calamite fruit, and Dr. Goeppert's on *Aphyllostachys*, afforded the writer much information; and he obtained the evidence of *Volkmannia* being the fruit of *Calamodendron* from the similarity of the central axes. About two years since he cleared up both the above points to his satisfaction, but Dr. Hooker had returned the specimens, and was prevented by press of business from joining him in publishing. The author then commenced the present Memoir unaided, and the plates were put in the engraver's hand. Owing to circumstances over which he had no control, the publication of the Memoir has been delayed longer than he expected.

In the mean time, in Memoirs referred to at page 10, Mr. Carruthers has described the Calamite and its fruit mainly from some of the Author's specimens above mentioned, which Dr. Hooker had lent him with others. The published observations are, of course, independent of each other; and, whatever may be their relative value, the Author wishes the dates and history of the specimens and investigations to be clearly stated.—E. W. B.—January 30th, 1868.

PLATE I.

Calamodendron commune.

Fig. 1 (No. 1).   A decorticated specimen from the Upper Brooksbottom Seam of Coal, Lancashire.   Natural size.

Fig. 2 (No. 2).   The furrowed and ribbed cast of the central axis of a large specimen from the same locality as No. 1.   Natural size.

In the following Plates the same parts in the specimens figured are indicated by the same letters, as follow :

a a. The middle part, showing the central axis, or pith, composed of large hexagonal utricles.

b b. The pseudo-vascular cylinder, composed of quadrangular tubes, having all their sides perforated by oval openings ; they are arranged in wedge-like bundles and in radiating series, originating from "orifices" near the thin end of the wedge next the central axis, increasing in size as they approach the circumference, and divided by medullary (?) bundles.

c c. Wedge-shaped masses of lax tissue, composed of oblong cells, in radiating series, having the thick end of the wedge next the central axis.   Both wedges and cells increase in size in the opposite direction to that seen in the tubes and masses of the pseudo-vascular cylinder, *b b.*

d. Diaphragm, apparently composed of cellular tissue, dividing the central axis or pith horizontally at the Joints.

e e. Oval-shaped bundles of pseudo-vascular tissue, passing from the central axis, or pith, and communicating with the Leaves.

f f. Bundles of cellular tissue, of an elongated oval form, near the circumference of the stem, dividing the pseudo-vascular cylinder, and having the appearance of medullary bundles, though only extensions of the wedge-shaped masses, *c c.*

g g. The central axis or column of a Cone.

h h. The Scales or Leaves forming the divisions of the Sporangium-receptacles or cells of a Cone.

i i. The Sporangium-bearers of a Cone.

j j. Sporangia full of Spore-like bodies.

k k. Portions of the bag containing the Sporangia.

Plate I

Fig 1.

Nº 1.

Fig 2.

Nº 2.

J. N. Fitch del et lith

R. Fitch imp

PLATE II.

Calamodendron commune.

Fig. 1.   Transverse section of Specimen No. 1.   Natural size.

Fig. 2.   A portion of the transverse section of the same specimen, showing parts of two wedge-shaped bundles of pseudo-vascular tissue, originating at two " orifices" on the outside of the central axis, and parts of the three wedge-shaped masses of coarse tissue, arranged in radiating series.   Magnified 10 diameters.

Fig. 3.   A longitudinal section of the same specimen, showing the walls of the tubes forming the pseudo-vascular cylinder perforated by elongate-oval openings; taken near to the central axis.   Magnified 70 diameters.

Fig. 4.   A tangential section similar to the last, taken nearer to the outside of the specimen, showing the oval openings on the walls of the tubes.   Magnified 70 diameters.

Fig. 5.   A tangential section of No. 1 Specimen, taken near the central axis, showing the pseudo-vascular bundles and the lax tissue dividing them.   Magnified 15 diameters.

Fig. 6.   A section similar to the last, taken near the middle of the specimen. Magnified 40 diameters.

Plate II

Fig. 1.

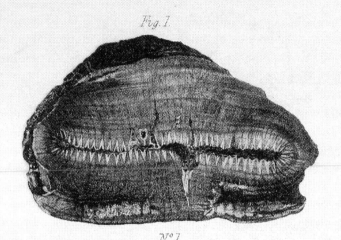

Nº 1

Fig. 4.

b
Nº 1.

Fig 2.

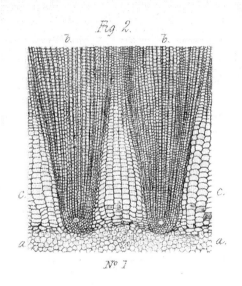

Nº 1

Fig 3.

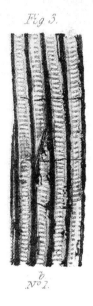

b
Nº 1.

Fig 6.

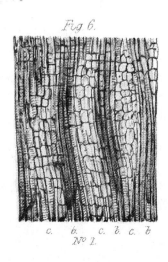

c. b. c. b. c. b
Nº 1.

Fig 5

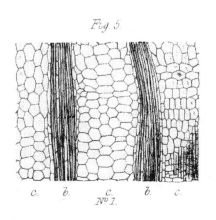

c. b. c. b. c.
Nº 1.

J.N.Fitch, del et. lith.

R. Fitch, imp

PLATE III.

Calamodendron commune.

Fig. 1. Specimen No. 3, from the " Hard Seam" of Coal at South Owram, near Halifax; showing the outside of a decorticated plant, of small size, with ribs, furrows, and a joint. Magnified 2 diameters.

Fig. 2. Transverse section of the same specimen, showing the wedge-shaped bundles of pseudo-vascular tissue, originating from "orifices" and parted by lax tissue. The central axis has for the most part disappeared. Magnified 10 diameters.

Fig. 3. A longitudinal section of the same stem, showing one of the horizontal diaphragms. Magnified 7 diameters.

Fig. 4. A tangential section of the same specimen, showing three oval-shaped bundles of vessels found at the joint, as well as some elongated oval masses of cellular tissue, not unlike medullary bundles. Magnified 12 diameters.

Fig. 5. A longitudinal section of a portion of the same specimen, showing the oval openings on the walls of the tubes forming the pseudo-vascular cylinder. Magnified 50 diameters.

Fig. 6. A tangential section of the same specimen, showing one of the medullary (?) bundles. Magnified 45 diameters.

Fig. 7. Transverse section of a very young stem (Specimen No. 4), slightly compressed, showing the structure of the central axis, the wedge-shaped bundles of pseudo-vascular tissue originating from "orifices," and parted by coarse tissue. From the Upper Brooksbottom Seam of Coal. Magnified 34 diameters.

Plate III.

Fig. 6.

Fig. 1.

Fig. 5.

Nº 3.

Nº 3

Nº 3.

Fig. 4.

Fig 2.

Fig 3

Nº 3

Nº 3

Nº 3.

Fig 7

Nº 4.

J N Fitch del et lith.

F Fitch imp.

PLATE IV.

Calamodendron commune.

The specimens in this Plate are from the Upper Brooksbottom Seam of Coal.

Fig. 1 (No. 5).   Transverse section of one of the Cones or organs of fructification, showing the central axis, composed of hexagonal and pentagonal vessels, and twelve Sporangia, full of Spore-like bodies ; the four of the lower ones enclosed in two cordate bags.   Magnified 45 diameters.

Fig. 2 (No. 6).   Transverse section of another Cone, showing the central axis and six processes or spines (Sporangium-bearers) radiating therefrom, and disarranged Spore-cases, formerly connected with them.   Magnified 56 diameters.

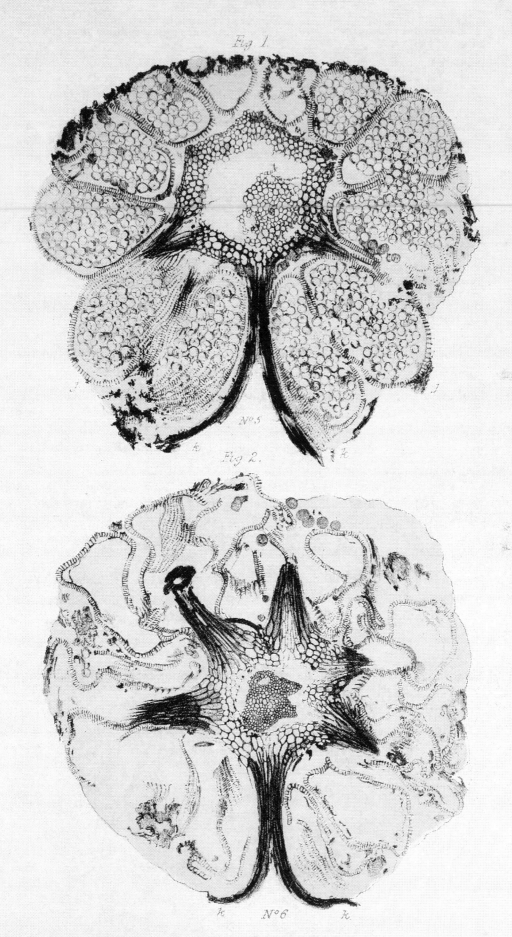

Fig. 1.

No 5

Fig. 2.

No 6

PLATE V.

Calamodendron commune.

All the specimens in this Plate are from the Upper Brooksbottom Seam of Coal.

Fig. 1 (No. 7). Longitudinal section of a Cone, showing seven Receptacles, Scales, and Sporangium-bearers, connected with the column or central axis, and Sporangia full of Spore-like bodies. Magnified $13\frac{1}{2}$ diameters.

Fig. 2 (No. 8). Longitudinal section of a Cone similar to the last, showing eight Receptacles and Sporangium-bearers, connected with the column or central axis, and Sporangia containing Spore-like bodies. Magnified 14 diameters.

Fig. 3 (No. 9). Longitudinal section of a single Receptacle with its two scales; and a tangential section of four cordate Sporangia, two of them containing dark coloured Spore-like bodies. Magnified 19 diameters.

Fig. 4 (No. 10). A longitudinal section of part of another Cone, showing the structure of the Scales and Sporangium-bearers, and the connection of both those parts with the column or central axis. Magnified 24 diameters.

Fig. 5 (No. 11). A longitudinal section of part of another Cone, showing the Scales and Sporangium-bearers, connected with the central axis or column, and the Sporangia, in their natural position, full of Spore-like bodies. Magnified 18 diameters.

Fig. 5*a* (No. 11). A longitudinal section of a single Sporangium full of Spore-like bodies, taken from No. 11 Specimen.

Fig. 5*b* (No. 11). A longitudinal section of a portion of the central axis or column of No. 11, showing the oval openings in the walls of the tubes. Magnified 130 diameters.

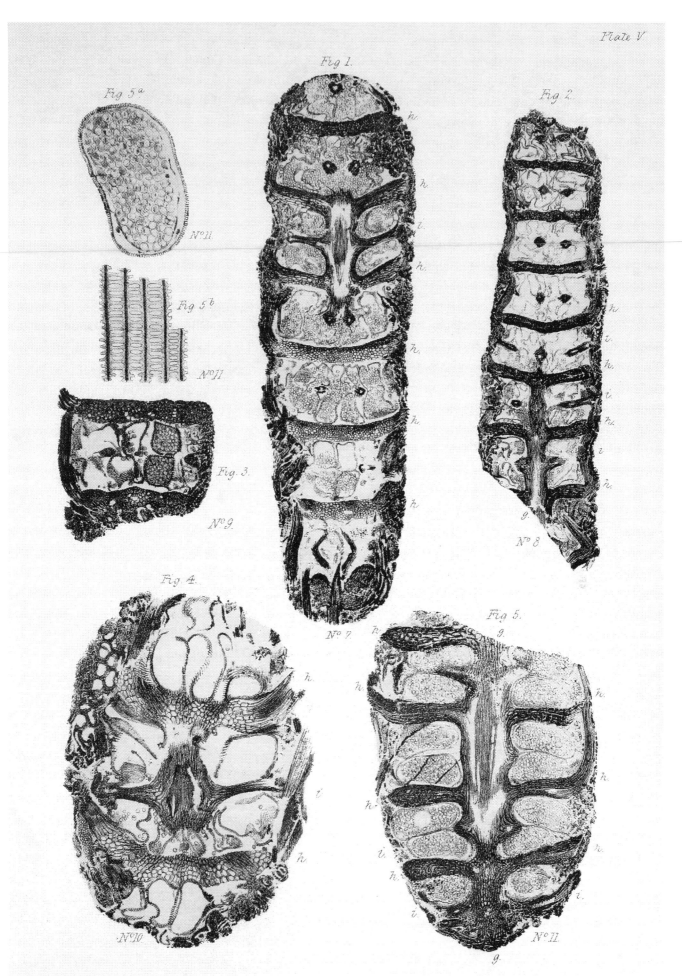

Plate V

Fig. 1.

Fig 5ª

Fig 2

Nº 11

Fig 5ᵇ

Nº 11

Fig. 3.

Nº 9.

Nº 7.

Nº 8.

Fig. 4.

Fig 5.

Nº 10.

Nº 11.

PLATE VI.

Fig. 1 (No. 12). Fruit-stalk, with Cones attached, resembling the *Aphyllostachys Jugleriana* of Goeppert; from the strata adjoining the Lower Brooksbottom Seam of Coal, at Ewood Bridge, Lancashire. Magnified half as large again as the original.

Fig. 2 (No. 13). Fruit-stalk, with Cones attached, resembling *Volkmannia sessilis* of Presl; from the Mountain-limestone at Holywell, North Wales. Magnified half as large again as the original.

Fig. 3 (No. 14). Specimen of *Asterophyllites longifolia*, from the Upper Coal-measures at Ardwick, near Manchester. Magnified half as large again as the original.

Fig. 4 (No. 15). Fruit-stalk of a Plant resembling *Calamodendron commune* (?), with Cones and Leaves attached to it, from the Upper Coal-measures at Ardwick, near Manchester. Magnified twice the natural size.

Fig. 4a (No. 16). Specimen showing six Scales, with their apices, attached to the central axis of a Cone, similar to those last described. From the Upper Coal-measures at Ardwick. Magnified 3 diameters.

E. W. B.

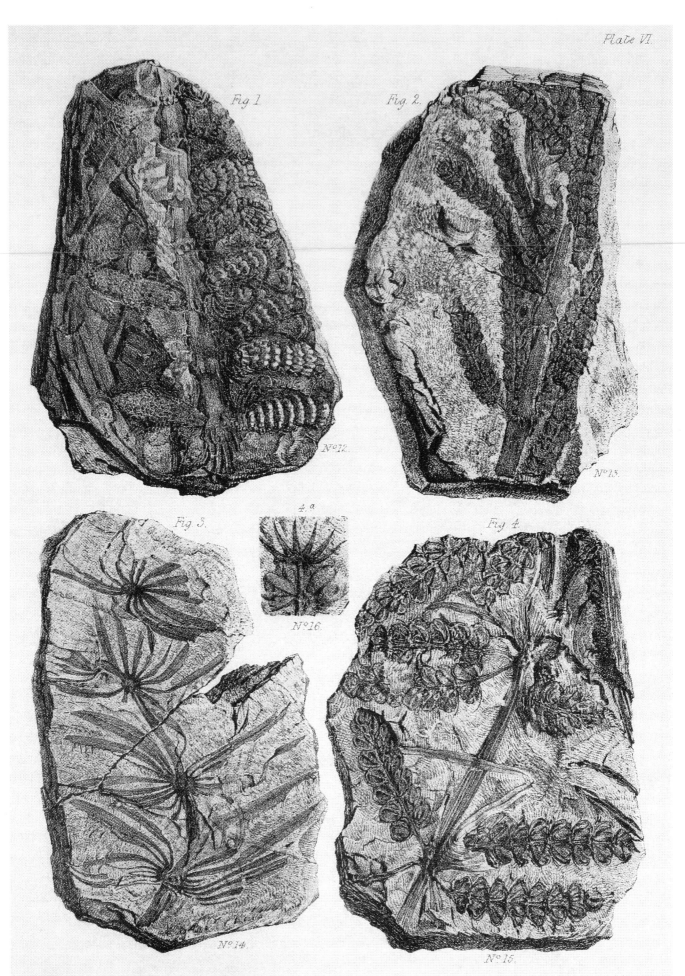

Plate VI.

Fig 1.

Fig. 2.

Nº12.

Nº13.

Fig 3.

#.a

Nº16.

Fig 4.

Nº14.

Nº15.

THE

PALÆONTOGRAPHICAL SOCIETY.

INSTITUTED MDCCCXLVII.

VOLUME FOR 1870.

LONDON:

MDCCCLXXI.

OBSERVATIONS

ON THE

STRUCTURE OF FOSSIL PLANTS

FOUND IN THE

CARBONIFEROUS STRATA.

BY

E. W. BINNEY, F.R.S., F.G.S.

PART II.

LEPIDOSTROBUS AND SOME ALLIED CONES.

PAGES 33—62; PLATES VII—XII.

LONDON:

PRINTED FOR THE PALÆONTOGRAPHICAL SOCIETY.

1871.

PRINTED BY

J. E. ADLARD, BARTHOLOMEW CLOSE.

CONTENTS.

                                                                    PAGE

I. INTRODUCTORY REMARKS 33

II. BIBLIOGRAPHICAL HISTORY OF LEPIDOSTROBUS AND FLEMINGITES ; WITH REMARKS ON THEIR
    RELATIONSHIP TO LEPIDODENDRON AND SIGILLARIA 33

    § 1. General Remarks 33

      2. *Lepidodendron.* Stem 34

      3. — Internal structure ; Witham and others . . . 35

      4. — — Hooker 35

      5. — Roots, Bowman, and Binney 36

      6. — — Richard Brown 36

      7. — — Hooker 36

      8. — — Carruthers 37

      9. — — — and Binney 37

     10. *Lepidostrobus.* Lindley and others 38

     11. — Robert Brown 38

     12. — Hooker 39

     13. — Carruthers 39

     14. *Flemingites.* Morris 40

     15. — Binney 41

     16. — Balfour 41

     17. — Goldenburg 42

     18. — Binney 42

     19. — Carruthers 42

     20. — Brongniart 42

     21. — Schimper 43

III. REMARKS ON MACROSPORES AND MICROSPORES 44

CONTENTS.

		PAGE
IV. DESCRIPTION OF THE SPECIMENS		46
§ 1. Specimens Nos. 17, 18, 19, and 20		46
No. 17. *Lepidodendron Harcourtii*		46
No. 18. — —		48
No. 19. — *vasculare*		49
No. 20. — —		50
2. Specimens Nos. 21 and 22		51
No. 21. *Lepidostrobus Russellianus*		51
No. 22. — —		52
3. No. 23. — (?) *dubius*		52
4. No. 24. — *tenuis*		53
5. No. 25. — *levidensis*		54
6. No. 26. — *Hibbertianus*		55
7. No. 27. — (?) *ambiguus*		55
8. No. 28. — *Wuenschianus*		56
9. No. 29. — *latus*		57
10. No. 30. *Bowmanites Cambrensis*		59
V. CONCLUDING REMARKS		60

PART II.

I. Introductory Remarks.

In this Monograph it was originally intended to have described a series of fossil Stems, showing structure, from the Coal-measures; and the genus *Calamodendron* was first taken as yielding, from specimens in my possession, more complete materials, not only for illustrating the structure of the stem, but also of the organs of fructification. Before proceeding to describe the structure of more stems, it has been considered desirable to bring before the Public some specimens showing Organs of Fructification. These are not very numerous, but their publication may be of use in indicating how small the amount of knowledge on the subject we really possess as yet, and in directing the attention of Collectors to the matter; for a great deal of labour is undoubtedly yet required before we shall be able thoroughly to understand the true nature of some of our commonest Coal Plants, so as to restore them in all their parts. One good specimen showing the organs of fructification connected with the stem and foliage of the plant is worth any number of detached fragments. It has been my good fortune to become possessed of a specimen showing such three portions of a plant; and, therefore, it has occurred to me that no time should be lost in describing it, although in due order, probably, it ought to have been delayed to a later portion of the Monograph.

II. Bibliographical History of *Lepidostrobus* and *Flemingites*, with Remarks on their Relationship with *Lepidodendron, Sigillaria*, &c.

§ 1. *General Remarks.*—The genera *Lepidodendron, Lepidostrobus,* and *Flemingites,* especially the two former, have been often described, in most works that have treated on Fossil Botany. The specimens figured have been for the most part mere casts and impressions, not showing any trace of internal structure. Those instances where the tissues of the plant happened to be well preserved, or where the organs of fructification distinctly showed their spores, have not been numerous and are soon given.

The specimen of *Lepidodendron Harcourtii* found by the Rev. C. G. W. Harcourt, Rector of Rothbury, and preserved by Professor John Phillips, afforded materials for the memoirs of Mr. Witham,[1] Messrs. Lindley and Hutton,[2] and M. Adolphe Brongniart,[3] on the structure of the stem.

Professor Morris[4] first described the capsules, so generally found in " splint " or " bone " coal, on which Mr. Carruthers[5] has established his genus *Flemingites*.

Dr. Hooker[6] collected materials which enabled him to show the anatomy of *Lepidostrobus*.

Dr. Robert Brown,[7] in a memoir read before the Linnean Society in 1847 and published in 1851, gave to the world a more perfect specimen of a portion of a cone than had been previously figured.

Mr. Carruthers[8] has shown what he considered to be the differences between *Lepidostrobus* and his new genus *Flemingites*.

And latterly M. Adolphe Brongniart[9] and Professor Schimper[10] have obtained specimens of the most perfect cones, exhibiting both microspores and macrospores in the same cone.

§ 2. *Stem.*—For many years the beautiful stems of *Lepidodendron* have attracted the attention of collectors of Coal-plants, on account of the elegant sculpture of the spots where the leaf proceeded from the stem. In the ordinary condition of the fossil plant, whether found in shales or sandstones, these have the appearance of scales, and hence the name. But this was not the original form of the scar of attachment, as Dr. Hooker[11] first showed; for the pitted knob, afterwards compressed into the so-called scale, projected originally from the stem one fourth to one third of an inch, as specimens in my cabinet prove. Different " species " have been named, owing to specimens being more or less compressed out of their original form. M. Ad. Brongniart has enumerated a great number of so-called species, which Professor Schimper has reduced to fifty-six. Doubtless these will have to be still further reduced, as in the case also of *Sigillaria*, when we become

[1] 'Observations on Fossil Vegetables,' by Henry Witham, 4to, Edinburgh, 1831; and 'On the Internal Structure of Fossil Vegetables found in the Carboniferous and Oolitic Deposits of Great Britain,' by H. T. Witham, of Lartington, 4to, Edinburgh, 1835.

[2] 'The Fossil Flora of Great Britain,' by Dr. Lindley and W. Hutton, 3 vols. 8vo, London, 1831-37.

[3] 'Histoire des Végétaux fossiles, ou Recherches Botaniques et Géologiques,' &c., 2 vols. 4to, Paris, 1828.

[4] 'Transact. Geol. Soc. London,' 2nd series, vol. v (explanation of plate xxxviii), 4to, London, 1840.

[5] and [8] 'Geological Magazine,' vol. ii, p. 433, 8vo, London, 1865; and ibid., vol. vi, pp. 151, &c., 1869.

[6] 'Memoirs of the Geological Survey of Great Britain,' vol. ii, part 2, pp. 440, &c., 8vo, London, 1848.

[7] 'Transact. Linnean Society,' vol. xx, 4to, London, 1851.

[9] 'Comptes Rendus,' vol. lxvii, pp. 421, &c., &c., 4to, Paris, 1868.

[10] 'Traité de Paléontologie végétale,' vol. ii, p. 69. Paris, 1870.

[11] 'Mem. Geol. Surv.,' vol. ii, part 2, p. 444.

better acquainted with their true original external character before the plants had been subjected to pressure. *Sigillaria,* instead of the flat surface and leaf-scars generally seen, had, in large specimens, strong ribs projecting fully an inch beyond the furrows, and divided by narrow furrows.

§ 3. *Internal Structure (Witham and others).*—For the first knowledge of the internal structure of the stem of *Lepidodendron* we are indebted to Mr. Witham,[1] Messrs. Lindley and Hutton,[2] and M. Adolphe Brongniart.[3] The branches dichotomised with great regularity, and at their extremities produced the cone so well known as *Lepidostrobus.* This latter fact had, for the most part, been inferred from the number of specimens of *Lepidostrobi* being found around and about the stems of *Lepidodendron,* rather than by any actual proofs of the cone being found actually attached to the stem itself.

§ 4. *Internal Structure and Cones (Hooker).*—Dr. Hooker, to whom we are indebted for most valuable information as to the nature of *Lepidodendron,* in his memoir previously quoted gives, at page 451, the following descriptions :

" 1. LEPIDODENDRON *elegans.* Cone slender, three quarters of an inch in diameter, four to ten inches long ; sporangia eight in a whorl.

" 2. LEPIDODENDRON *Harcourtii ?* Cone broad, one and a half inch in diameter ; sporangia about sixteen in a whorl.

" If, now, these cones be examined with reference to the known contemporaneous fossils which accompany them, it will appear impossible to deny their having been the reproductive organs of *Lepidodendron,* not only from their association with the fragments of that genus, because the arrangement of the tissue in the axis of the cone entirely accords with that of the stem of *Lepidodendron,* just as we find in modern cones of *Lycopodiaceæ* and *Coniferæ* that the axis is a continuation of the branch which bears leaves, modified into organs adapted to support and protect the parts of fructification."

Now, although the author was quite right in his conclusion that *Lepidostrobus* was the fruit of *Lepidodendron,* in the beautiful plates which accompany his memoir there is no good transverse section given to prove the fact; and some writers, especially Mr. Carruthers, consider the evidence of the connection between a *Lepidodendron* and its own *Lepidostrobus* to be consequently of a very unsatisfactory nature.[4] Of course the occurrence of the cones in the insides of stems is by no means satisfactory, as any one knows who has examined the stems of *Sigillaria,* wherein are found plenty of Ferns, which certainly would not be admitted

[1] 'On the Internal Structure of Fossil Vegetables found in the Carboniferous and Oolitic Deposits of Great Britain,' by H. T. Witham, Edinburgh, 1833.

[2] 'Fossil Flora of Great Britain,' vol. ii, p. 46.

[3] 'Hist. Végét. foss.,' vol. ii, pp. 37 and following, plates x and xi.

[4] "On an Undescribed Cone from the Carboniferous Beds of Airdrie, Lanarkshire," by W. Carruthers, F.L.S. ; 'Geol. Mag.,' vol. ii (No. xvi), p. 437.

to be the foliage of that plant merely from being found in the interior of its stem. The same may be said of *Stigmariæ* found in nodules in the Wigan Four Feet Coal, which are full of spores, but up to this time, to my knowledge, no one has clearly proved such spores to belong to *Sigillaria*.

In pl. v, figs. 10, 11, 12, and 13, of Dr. Hooker's memoir, there is, in my opinion, clear evidence to prove that the structure of the stem of *Lepidodendron* is identical with that of a *Lepidostrobus*. In this Monograph proof will be adduced in confirmation of Dr. Hooker's opinion, by showing that the stem of *Lepidodendron Harcourtii* is identical in structure with the axis of *Triplosporites* or *Lepidostrobus Brownii* of Carruthers, and that the stem of *Lepidodendron vasculare* had a cone resembling that of *Lepidodendron Harcourtii* in the form of its sporangia, with the bracts and leaves also similar, but was still different in the structure of its central column.

§ 5. *Roots (Bowman and Binney).*—As to the roots of *Lepidodendron*, we are not in possession of much definite information. In the remarkable fossil trees described by the late Mr. J. E. Bowman, F.G.S.,[1] and which I had the pleasure of visiting with my lamented friend, the specimens were clearly of two kinds, most of them being *Sigillaria*, but certainly one or two of them having markings on their exterior like those usually found on *Lepidodendron*. Around and at the bases of the stems were nodules of clay-ironstone, in the shales, all full of *Lepidostrobus*, chiefly *L. variabilis*. At that time both of us believed that such cones had been connected with, and in fact had constituted, the fruit of the stems, but of which of them we did not hazard an opinion.

§ 6. *Roots (Rd. Brown).*—Mr. Richard Brown, in his description[2] of an upright *Lepidodendron* with *Stigmarian* roots, in the roof of the Sydney Main Coal, in the Island of Cape Breton, gives very good evidence of the root of his specimen being *Stigmaria* ; but as to the stem being a *Lepidodendron*, it does not appear so certain from his published illustrations, at least the evidence adduced cannot by any means be considered conclusive as to whether the specimen belonged to *Lepidodendron* or to *Sigillaria*.

§ 7. *Roots (Hooker).*—Dr. Hooker, in his " Vegetation of the Carboniferous Period, as compared with that of the Present Day,"[3] says, " Of the stems, branches, leaves, and fructification [of *Lepidodendron*] we have thus a very satisfactory knowledge, but the nature of their roots is not ascertained. Mr. Dawes, of West Bromwich, to whom I am indebted for much information regarding the structural characters of Coal Fossils, is inclined to regard the species of *Halonia* as roots of *Lepidodendron*, on which opinion I have no remarks to offer."

[1] 'Transactions of the Manchester Geological Society,' vol. i, pp. 112—184, 8vo, 1841.

[2] 'Quarterly Journal of the Geological Society of London, vol. iv, pp. 46, &c., 1847.

[3] 'Mem. Geol. Surv.,' vol. ii, part 2, p. 422.

§ 8. *Roots* (*Carruthers*).—Mr. Carruthers, in his paper on the structure and affinities of *Lepidodendron* and *Calamites*, published[1] in 1867 (at page 5 of the excerpt), says, "Stigmarioid roots have been determined to belong to *Lepidodendron*, as well as to *Sigillaria*, and their whole structure supports this determination. I have satisfied myself that there is nothing that can be truly called a medullary ray in the woody cylinder of *Stigmaria*, but into the proof of this I will not now stay to enter. The base of the trunk was divided into a few principal roots, and these again divided dichotomously, but the ultimate divisions were never much attenuated. Throughout their whole course, and from every portion of their circumference, they gave off rootlets of considerable length, which, with the exception of a slender vascular bundle, were entirely composed of delicate hexagonal cells. They were articulated to flagon-shaped bodies, sunk in cavities, arranged in a quincuncial manner over the stem. The internal structure of *Stigmaria* corresponds to that of the trunk of a *Lepidodendron*. The axis was composed of fusiform barred cells, and this was surrounded by a woody cylinder, which was certainly penetrated by the vascular bundles that supported the rootlets. Beyond the woody cylinder came a great thickness of cellular tissue, almost always destroyed, but probably agreeing in its structure with the three zones of the stem."

§ 9. *Roots* (*Carruthers and Binney*).—As Mr. Carruthers' conclusions, which really go to the extent of proving that *Sigillaria* and *Lepidodendron* are the same plant, appear to be drawn mainly from my specimens of *Sigillaria vascularis* and *Lepidodendron vasculare*, described in the 'Quarterly Journal of the Geological Society,' vol. xviii, pp. 106—112, pls. iv, v, and vi, for May, 1862, it is desirable carefully to consider the subject. Of course the specimens had externally all the usual characters of *Lepidodendron*, but in the internal structure No. 3 differed considerably from Nos. 1 and 2, in possessing no internal radiating cylinder outside the medullary sheath, and it was, therefore, classed in the genus *Lepidodendron*. Up to the present time no evidence has come to my knowledge to show that the central axis of *Stigmaria ficoides* was the same as that of *Lepidodendron vasculare;* indeed, we are really ignorant of the structure of the medulla of *Stigmaria*, except so far as Professor Goeppert has made us acquainted with it, and as far as my own specimens of *Sigillaria* show ; but it is no doubt the same as that of *Sigillaria vascularis*. This is probably not sufficient to warrant us in saying more at present than that *Sigillaria* and *Lepidodendron* were very much alike, and had similar habitats, and it is scarcely sufficient to prove that they were one and the same plant.

Unquestionably the medulla of the *Lepidodendron Harcourtii* of Witham and Lindley and Hutton, as well as a much more perfect specimen of the same species in my cabinet, presented to me by Mr. J. S. Dawes, and hereinafter described, is composed of cellular tissue, without any vascular bundles dispersed through it ; whilst in my *Sigillaria vascularis*, both in the small specimens with rhomboidal scars and the large irregularly ribbed

---

[1] 'In the Journal of Botany,' 8vo, London, 1867.

and furrowed examples, as well as in my *Lepidodendron vasculare*, the medulla contains numerous scalariform tubes or utricles, called by Mr. Carruthers "fusiform barred cells." In addition, we must not forget that in *Lepidodendron Harcourtii* we find no trace of the internal radiating cylinder so characteristic of *Sigillaria vascularis*, although in both plants, as well as in *Lepidodendron vasculare*, there is evidence of an outer radiating cylinder composed of fusiform utricles. This external cylinder has been shown by me to occur in *Stigmaria*.[1]

In the present Monograph there will be described and figured two cones showing the internal composition of their columns; one of these exactly resembles in its structure that of *Lepidodendron Harcourtii*, and the other is similar to that of *Lepidodendron vasculare*; so for the present, probably, it will be desirable to keep the two genera distinct.

So far as my observations extend, plenty of both large and small specimens of *Lepidodendron* having rhomboidal scars are met with, whilst few ribbed and furrowed *Sigillariæ* of a small size are found, these fluted *Sigillariæ* being for the most part large specimens. In my memoir last quoted it was shown that, while the large stem of *Sigillaria* had small branches rhomboidally scarred and of a lepidodendroid character, the large stems, having the same internal structure as the branches, resembled the ordinary *Sigillaria* in its external appearance.

Hence, although *Lepidodendron* and *Sigillaria* are, no doubt, nearly allied, it appears to me desirable to retain them as distinct genera for the present, until more evidence of their perfect identity is furnished than my specimens figured and described in the 'Quarterly Geological Journal' and the 'Philosophical Transactions' have afforded.

LEPIDOSTROBUS.

§ 10. (*Lindley and others*).—From the earliest time that Carboniferous fossil Plants have been figured and described, the *Lepidostrobus* has appeared amongst them. Messrs. Lindley and Hutton, in their 'Fossil Flora,' give numerous figures of beautiful specimens, as also does M. Brongniart in his 'Histoire des Végéteaux fossiles;' but none of these appeared to show any internal structure, owing to their being simply casts, or carbonaceous impressions, in clay-ironstone or shale.

§ 11. (*Brown*).—Dr. Robert Brown read before the Linnean Society in 1847 a description of a cone (*Triplosporites*) which first showed the true structure of *Lepidostrobus*, so far, at least, as the upper part of the cone was concerned, but his paper was not published till 1851. In the mean time, namely in 1848, Dr. Hooker published his valuable researches on the structure of *Lepidostrobus*, showing the true nature of the cone, from a number of

---

[1] 'Philosophical Transactions,' vol. clv (1865), p. 593.

specimens, none of which were so perfect as that of Dr. Brown's, which exhibited the sporangia of the upper part of the cone, full of microspores, and attached to the central column, as well as the form of the microspores themselves. Both these authors came to the same conclusion from independent specimens—Dr. Brown from one in a perfect state of preservation, so far as the upper portion of the cone was concerned, and Dr. Hooker from several imperfect specimens, taking a little from one and a little from another. But neither of these observers was so fortunate as to see the lower part of *Lepidostrobus* containing sporangia with spores in them.

§ 12. (*Hooker*).—The genus *Lepidostrobus* has long been considered the fructification of *Lepidodendron*. This must of necessity have been the case as soon as *Lepidodendron* was supposed to be allied to *Lycopodium* by M. Brongniart and Dr. Lindley.

Dr. Hooker, in his excellent memoir on *Lepidostrobus*, published in the 'Memoirs of the Geological Society,'[1] notices numerous specimens of *Lepidostrobus* in the insides of stems of *Lepidodendron Harcourtii* and *L. elegans*. He says, at page 451, "The two cones from which the above general views have been deduced are apparently from different species, and I shall therefore characterise them as such. Further, as they seem to have belonged to the fossil *Lepidodendron* enclosing them, I shall for the future notice them as really the fruit of that genus." He then describes the cone of *Lepidodendron elegans* as slender, three quarters of an inch in diameter, four to ten inches long, with sporangia eight in a whorl; and the cone of *L. Harcourtii* (?) as broad, one inch and a half in diameter, with sporangia about sixteen in a whorl. He observes, "The most positive evidence that can be adduced of *Lepidostrobi* being a genus allied to *Lycopodium* is afforded by the spores, the presence of which not only removes them from *Cycadeæ*, *Coniferæ*, or any other order of flowering plants, but directly refers them to the family of *Lycopodiaceæ* and [not to ?] *Coniferæ*. In both [*Lepidostrobus* and *Lycopodium*] the original spore divides into three or nearly [rarely ?] four sporules, which are angular, and form the reproductive system of the plant. Not only do these groups coincide in the essential characters of their spores, but in many minor points the strongest similarity exists between them. The arrangement of the scales is the same in both, and so are the scales themselves in general features, especially towards their dilated apices. The situation of the sporangia, too, is alike, and their attachment by a very narrow surface to the scale."

§ 13. (*Carruthers*).—Mr. Carruthers, speaking of *Lepidodendron*, says,[2] "The fruit was a strobilus (pl. lvi, fig. 3) formed from a shortened branch, the leaves of which are converted into scales, that support on their upper surface a single large sporangium

[1] Vol. ii, part 2, 1848.

[2] "On the Structure and Affinities of Lepidodendron and Calamites," 'Journ. Botan.' vol. xiii, pp. 6 and 7.

(*Lepidostrobus*, pl. lvi, fig. 4), or perhaps several small ones (*Flemingites*, pl. lvi, fig. 6).

"There appear to be both macrospores (pl. lvi, fig. 5) and microspores in the same sporangium. I have examined at length the structure and affinities of these fruits, in a paper published in the 'Geological Magazine,' vol. ii, p. 433, to which I must refer, without here dwelling further on the subject. *Flemingites*, although the sporangia are enormously abundant in some coals, have not yet been found connected with any fossil; but specimens of *Lepidostrobus* attached to branches of *Lepidodendron* have been described by Dr. Paterson, Brongniart, and others, and I have noticed a fine specimen in the Museum of the Edinburgh Botanic Gardens, and others exist in the British Museum and elsewhere.

"In tracing the affinities of *Lepidodendron*, we have the safest guide in the organs of fructification, and fortunately these have been satisfactorily determined. The sporiferous strobilus shows that it is a true cryptogam, and in general appearance and arrangement of parts the strobilus can scarcely be distinguished from that of some living *Lycopodia*, except in the great difference of size. This affinity is strengthened by the character of the leaves and the structure of the stem. But the possession of both kinds of spores in the same sporangium exhibits stronger affinity to *Rhizocarpeæ* than to *Lycopodiaceæ*."

FLEMINGITES.

§ 14. (*Morris*).—Probably ever since the Splint Coals of Scotland have been wrought the small disc-shaped bodies, of a chestnut colour, contained in them must have been observed, but until late years little attention has been devoted to them. They have been found in Cannel and Cherry as well as in Splint Coals, but in nothing like the same quantity. In some of the Fifeshire "Splints," near Methel, they occur in such abundance as to impart to the coal a chestnut tinge.

In Mr. Prestwich's memoir on the Coalbrookdale Coal-field, Professor Morris noticed these bodies in the following words : [1]—

"*Lycopodites* [2]? *longibracteatus*, n.s. Stem rounded, marked by the cicatrices of the fallen leaves, which are close, lozenge-shaped, and spirally disposed.

"Fructification in terminal imbricated spikes; thecæ reniform, minutely tuberculated, each attached by its centre to a base of a *long, lanceolate, foliaceous* bractea.

"The thecæ resemble in shape those of the recent genus *Stachygynandrum*; but as

---

[1] 'Transact. Geol. Soc. London,' 2nd ser., vol. v, part 3 (1840), explanation of plate xxxviii, figs. 8, 9, 10, and 11.

[2] Professor Morris, I believe, has since determined this plant to be a true *Lepidodendron*.

the capsules vary in form in different parts of the spike in that genus, it is difficult to assign its affinity to that division of the *Lycopodiaceæ*.

" The capsules (pl. xxxviii, figs. 8 and 8 *a*) of this species neither bituminized nor mineralized, but in a state of brown vegetable matter, are very abundant in some of the coarser sandstones of the Coal-measures."

§ 15. *Binney.*—In a paper of my own, published in ' The Quarterly Journal of the Geological Society '[1] for May, 1849, a description is given of some spores found imbedded in the roots of *Sigillaria*, from a coal-seam at Wigan. These bodies varied in size from one twenty-fifth to one twentieth of an inch in diameter, were of a nearly spherical form, and had a tri-radiate ridge on the under portion. On carefully comparing them with the spores of *Lepidostrobi*, figured in plates 5, 6, 8, and 10, of Vol. ii, Part 2, of the ' Memoirs of the Geological Survey of Great Britain ' (Dr. Hooker's illustrations), I became acquainted with their true nature, and, from their resemblance, was led to believe them to be the spores of *Lepidostrobus ornatus*.

§ 16. *Balfour.*—Dr. Balfour, in a paper ' On certain Organisms found in Coal from Fordel,"[2] says, "Besides Sigillarias and Stigmarias, we also detect in the Fordel coal peculiar organisms, which have the appearance of seeds (pl. 11, figs. 12 and 13). Dr. Fleming informs me that similar bodies have been observed by him in coal, and that he exhibited them to Mr. Witham about twenty years ago. They have also been seen by Dr. Fleming in Lochgelly and Arniston parrot, and in the coal at Boghead; and, from having observed them in cherry, splint, and cannel coals, he is disposed to consider them as a somewhat common feature. I have seen them from coal at Miller Hill, near Dalkeith, as well as in the coal from Fife. They do not appear to have been fully described. The nearest approach to them is the *Lycopodites* figured by Mr. Morris in the ' Appendix to Mr. Prestwich's Paper on the Geology of Coalbrookdale.' They appear to be certainly allied to the fructification of the *Lycopodiaceæ* of the present day, more particularly to that form of it which consists of two valves placed in apposition, and containing what is called Lycopode powder, or minute cells having a glistening aspect, interspersed sometimes with matter of a dark wine-colour. These and like bodies I therefore consider to be the sporangia or spore-cases of some plant allied to *Lycopodium*, perhaps *Sigillaria*."

" Explanation of plate 11, figs. 5 to 18. Fig. 13. The same sporangia, magnified about eight diameters, imbedded in a mass of Fordel coal; some lying on the surface, others projecting from the broken edges of the coal. They seem to occur frequently in coal from different localities both in England and Scotland. Mr. Binney has seen them in Wigan coal."

[1] Vol. vi, p. 17, &c. ; with woodcuts, figs. 2, 3, 4.
[2] 'Transactions of the Royal Society of Edinburgh,' vol. xxi, p. 187.

§ 17. *Goldenberg.*—Dr. Goldenberg[1] describes and figures such spherical bodies, some with the triradiate ridge, and others without that character, as the fruit of *Sigillaria*, *Stigmaria*, and fossil *Selagineæ*. These bodies, according to this author, appeared to be attached to the scales of the cones, and not contained in a sporangium; and in the figures they appear chiefly at the base of the specimen.

§ 18. *Binney.*—In 1864 a description of some spores of plants found in the splint coals of Methel, Fifeshire, was given by myself to the Literary and Philosophical Society of Manchester,[2] wherein it was stated that, when we consider the great abundance of these small fossils in all splint coals, and the immense number of the roots of *Sigillaria* found in the floors of such seams of coal, it is almost certain that they had some connection with that plant. This tended to confirm M. Adolphe Brongniart's opinion, expressed many years ago, that *Sigillaria* and *Lepidodendron* were plants very nearly allied to each other.

§ 19. *Carruthers.*—In October, 1865, Mr. Carruthers described[3] a fossil cone that had been discovered by Mr. James Russell, of Chapelhall, Airdrie, a diligent and intelligent collector of Carboniferous Fossils, and which showed the bodies described by Professor Morris, and noticed by myself and Dr. Balfour, as they occurred in the strobilus, and not detached, as they had been observed by me and Dr. Balfour. Mr. Carruthers describes at length the differences he considered to exist between *Lepidostrobus* and his new genus *Flemingites.* The two genera are thus contrasted (p. 438):

"*Lepidostrobus.*—Each scale of the cone supporting a single oblong sporangium.

"*Flemingites.*—Each scale of the cone supporting a double series of roundish sporangia.

"*F. gracilis.*—Cone slender, cylindrical, very slightly tapering at the base, composed of a solid axis and numerous imbricated scales, ten in a whorl. The apex of the scale long and slender. Sporangia attached by a tri-radiate ridge."

§ 20. *Brongniart.*—M. Adolphe Brongniart in a notice, "Sur un fruit de Lycopodiacées fossiles," in the 'Comptes Rendus' for August, 1868, gives a description of a cone very similar in its upper portion to that described by Dr. Robert Brown, which he names *Triplosporites Brownii* (*Lepidostrobus Brownii*, Carruthers).

This specimen shows sporangia containing microspores in the upper part of the cone, exactly like those in Dr. Brown's specimen; whilst in the lower portion of the same

[1] 'Flora Saræpontana fossilis;' plate B, figs. 18 to 25 (1855); plate x, figs. 1 and 2 (1857).

[2] 'Proceedings of the Literary and Philosophical Society of Manchester,' vol. iv, (for 1864), p. 45.

[3] "On an Undescribed Cone from the Carboniferous Beds of Airdrie, Lanarkshire," by W. Carruthers, F.L.S., of the British Museum; 'Geological Magazine,' vol. ii (No. xvi, October, 1865), pp. 433, &c.

cone there are macrospores in the sporangia, resembling those by Mr. Carruthers as the sporangia of *Flemingites*. This is the first instance that has come to my knowledge of a fossil cone containing both microspores and macrospores. It appears to have been found in the Drift deposits of the valley of Volpe, in Haute-Garonne, by M. Dabadie; but, if it is, as M. Brongniart asserts, identical in structure, so far as its upper part is concerned, with Dr. Brown's specimen, there is no doubt that it originally came from the Carboniferous strata.

The learned author says at page 424, " Cet épi présente donc, comme les Lycopodiacées des genres *Selaginella* et *Isoëtes*, des sporanges de deux natures, les uns, vers le sommet de l'épi, contenant des microspores, c'est à dire, des anthéridies; les autres, situés vers la base de l'épi, renfermant des macrospores ou spores germinatives.

" La forme et le mode d'insertion des sporanges, leur grand volume, le numbre considérable de macrospores qu'ils renferment, l'absence de toute trace de ligne de déhiscence régulière, font surtout ressembler ces organes à ceux des *Isoëtes;* mais dans ces plantes ces sporanges sont insérés sur la base même des feuilles qui naissent d'une tige très-courte et bulbiforme. Dans la plante fossile, au contraire, ces sporanges sont portés par des sortes de bractées, ou feuilles squamiformes, réunies en un épi, qui, comme ceux des *Selaginella*, terminait probablement les rameaux.

" Il y a donc là une combinaison particulière de caractères : sporanges analogues à ceux des *Isoëtes* réunis en un épi semblable à celui des Lycopodes et beaucoup plus grand."

M. Brongniart has been so kind as to forward to me a drawing of a sporangium and the macrospores contained in it, as well as of the microspores from his cone.

§ 21. *Schimper.*—Dr. Schimper[1] describes and figures the same fossil as M. Brongniart has treated of, and, terming it " *Lepidostrobus Dabadianus*, Sch., describes it as oblongo-cylindraceus, centim. $11\frac{1}{2}$ longus, in medio centim. 5 latus, extus cicatricibus tectus hexagonis millim. 6—8 latis, totidem altis, exacte contigiis, in medio tuberculo irregulari laminæ deciduæ residuo instructis, secundum ordinem $\frac{2}{27}$ dispositis; microsporis strobili dimidium superius occupantibus, illis præcedentis similibus; macrosporis sporangia dimidii inferioris tenentibus multo majoribus, sphæricis, tetraedri solum cacumen monstrantibus."

Dr. Schimper appears to distinguish the macrospores of his *Lepidostrobus Dabadianus* from the capsules described by Goldenberg; and he classes my Wigan specimens, which evidently are the uncompressed forms of the same fossil, found in splint coals, with that author's macrospore. Under the head of *Sigillaria* he says (p. 105):

[1] ' Traité de Paléontologie végétale, ou la Flore du Monde Primitif dans ses rapports avec les formations géologiques et la Flore du Monde actuel.' Part I, vol. ii, p. 69.

" B. Spicæ fructificatonis.

Sigillariostrobus, ' Sch. Atlas,' Plate LXVII, figs. 12, 24.

" Spicæ pedicellatæ strobiliformes oblongo et elongato-cylindraceæ, bracteis e basi ovato triangulari subito angustatæ, lanceolatæ, medio costatæ. Sporæ sporangio bracteæ basis lateri anteriori adfixo (incluso ?) inclusæ, magnæ (macrosporæ ?) et minores (microsporæ ?) tetraedræ.

" Les épis que je rapporte avec M. Goldenberg aux *Sigillaria* se distinguent facilement de ceux des *Lepidodendron* par leurs bractées, dont la base sporangiophore est insérée presque verticalement, au lieu de l'être horizontalement, comme dans ces dernièr. Le sporange occupe toute la largeur de la base de la bractée, et parait avoir été d'une consistance très tendre. Les spores sont de grandeur differente, des macrospores et des microspores ; les premières offrant un diamètre de 1, 1½, à 2 millimètres, les autres à peine celui de 1 millimètre (voy. notre planche, figs. 16, 20, 23). Les macrospores se rencontrent souvent en très-grande quantité dans les couches à *Sigillaria* et *Stigmaria* et quelquefois dans l'intérieur de ces troncs.

" Les épis eux-mêmes étaient fixes au tronc entre les coussinets foliaires, soit en suivant les séries droites (orthostiques, voy. notre planche, fig. 2*a*), soit en suivant les lignes obliques ou la spire fondamentale. Nons avons donné plus haut la description des cicatrices que ces épis ont laissées sur les troncs."

III. Remarks on Macrospores and Microspores.

Professor Morris many years ago remarked of the capsules from the Coalbrookdale Coal-field that they are neither mineralized nor bituminized (see above, p. 41), but in a state of brown vegetable matter. On finding similar bodies in some of the Low Moor Coals, he attributed the excellent qualities of those beds for the manufacture of iron to the presence of the spores. It is well known that the soft caking coal of Low Moor, called the " Better Bed," as well as the celebrated hard coal of Elsicar, Yorkshire, and all the Scotch splint coals, have been long prized for their iron-making qualities ; but the goodness of the latter may have arisen from their great power of sustaining weight in the furnace, and their freedom from sulphur, as well as from their containing any peculiar hydrocarbon derived from the spores. In the rich Boghead and Methel Cannels the spore is found, but not in such quantity by any means as in the splint coals. On making a section of either of these last-named coals for microscopic observation, and examining it under a three-quarters power, it appears as a dark-coloured vesicular mass, having its vesicles filled with a yellowish matter. In the celebrated Boghead Trial these were mistaken, by some of the witnesses, for the cellular structure of plants, and all the evidence of the chemical witnesses went to show that they had not been able to dissolve such yellow matter out of the coal by the most powerful solvents. When, however, the coal is subjected to distillation at a low temperature, this matter goes off as a yellowish vapour, which, on being condensed, forms crude paraffine oil. On afterwards examining the coke, it is found to be a black pulverulent mass, very porous, and containing numerous vesicles, from which the yellow matter, first seen under the microscope, had been expelled by the heat. Now, although at present we cannot account for the rich oil-producing qualities of these coals by the macrospores found in them, it is possible that the microspores of some cones, not far removed from *Flemingites* or *Lepidostrobus*, may have largely entered into their composition and produced it. These smaller bodies appear to have been preserved in coal, like the larger ones, without having undergone much alteration; but, owing to their smaller size, they have generally not attracted much notice.

As might have been anticipated from their great power of resisting decomposition, the organs of fructification of recent plants would be most likely to be preserved at the present day, so we find that such also are the portions of plants most completely preserved in the coal beds. This is most probably owing to the fatty oils and waxy substances which protect them, as well as to the presence of tannin. Baron Reichenbach first discovered paraffine in the wood of the Beech, whose leaves have on their outsides a thick coating of waxy matter; and the great quantity of paraffine found in brown cannel coals, such as those of Boghead and Methel, may be partly due to the waxy matter of the organs of fructification and leaves of the ancient plants, with which they were enabled then to resist moisture, as is now found to be the case in the Cabbage, Nasturtium, and other plants; the paraffine now found in the coal being in much the same state as it existed in the old plants, and very little altered.

The microspores contained in the upper sporangia of *Lepidostrobus Brownii* (Carruthers), which may now probably be regarded as the fructification of *Lepidodendron Harcourtii*, are very little altered, and appear like crude paraffine, and different in composition from the sporangia in which they are enclosed, and the column to which they are attached, both these being chiefly composed of carbonate of lime.

In Mr. Carruther's specimens of the genus *Flemingites*, the small round bodies which he terms sporangia (see above, p. 42), but which some of the first living authorities now consider to be macrospores, appear to consist so far as their outer covering is concerned, of paraffine or some similar hydrocarbon, whilst their insides contain bisulphide of iron or carbonate of lime, according to the nature of the matrix in which the fossil occurs. Generally the scale which supported them, according to Mr. Carruthers, or the sporangium that contained them, as well as the column of the cone to which they were attached, have been converted into coal, and lost all their structure. In other cases the last-named portions of the plant have been converted into carbonate of lime, while the sporangia or macrospores consist of paraffine. In my cabinet is a mass of macrospores, near a cone with sporangia, and these bodies, on light being transmitted through them, show in their interior granular bodies, in appearance like the coriaceous envelope of the macrospore in which they are enclosed.

Mr. Carruthers, in his Memoir, appears to class all the spore-like bodies found in coals and ironstones, whether they have a rough or smooth outside, or are furnished with a tri-radiate ridge on the lower part or not, as sporangia, under his new genus *Flemingites*. This appears to me to be going probably rather too far, in the present state of our knowledge. These bodies, be they sporangia or macrospores, appear to have been a common form of the organs of fructification during the Carboniferous Epoch; and, although doubtless some of the sporangia containing them were arranged spirally round the column of the cone, as in *Lepidostrobus*, others were arranged verticillately in whorls around the axis, as in the organs of fructification of *Calamodendron commune*. As to the latter, Mr. Carruthers appears to think, according to his statement in the ' Geological Magazine '[1] (vol. vi, p. 155), that none of them have been found so arranged. A specimen, however, will be described in this Monograph, not only showing this verticillate arrangement, but the connection of the cone with a stem bearing branches and leaves, hence the genus *Flemingites*, if it remain, will scarcely suffice to hold all the disc-shaped bodies described by Professor Morris, myself, Goldenberg, Balfour, and others. As to their all being macrospores, it appears to me there can be little doubt but there are macrospores not of one plant, but of many distinct plants.

IV. DESCRIPTION OF THE SPECIMENS.

§ 1. THE SPECIMENS (*Lepidodendron Harcourtii* and *Lepidodendron vasculare*), Nos. 17, 18, 19, *and* 20. Plates VII and VIII.

Specimen No. 17, *Lepidodendron Harcourtii* (Plate VII, figs. 1—5, 7—10).

Fig. 1 is a fragment of a cone, one and a half inches in length, one and a quarter inches across its major, and one inch across its minor axis. Although doubtless originally cylindrical, it has now somewhat of an oval section. The outside of the fossil, which is deprived of the upper portions of its scales or bracts, exhibits rhomboidal scars in every respect similar to those usually found on *Lepidostrobus ornatus;* and is most probably either the middle or the upper part of such a cone. It was found in a calcareous nodule from the Upper Foot Coal, near Oldham (marked with three asterisks in the section previously

[1] In this paper Mr. Carruthers states that *Lepidostrobus variabilis* is really a specimen of *Flemingites gracilis*.

given at p. 12), by Mr. James Whitaker, of Watershedding Bar, near Oldham, who has kindly allowed me to slice the specimen and examine it.

Figure 2 is a transverse section of the specimen, of the natural size, showing the central axis and sixteen irregularly pear-shaped sporangia, each about half an inch in length, radiating from it.

Figure 3 represents a longitudinal section (natural size) of the central axis and the scales or bracts, which have a spiral arrangement and support sporangia, on each side of it.

Figure 4 is the transverse section of the cone, magnified two diameters, showing its internal structure. The centre, originally composed of cellular tissue, and for the most part destroyed, has been replaced with carbonate of lime. It is surrounded by a zone of hexagonal tubes, having their sides barred by transverse striæ. This is bounded on the outside by a sinuous dark line, from which are seen to spring the bundles of barred vessels that communicate with the scales or bracts. These vessels traverse, in a highly inclined curve upwards, a band of cellular tissue which has been mostly removed and replaced by mineral matter. The outside of the axis shows elongated cells or utricles arranged in radiating series, similar to what are usually found on the outside of stems of *Lepidodendron Harcourtii*, and described at length by Messrs. Witham, Lindley and Hutton, and Adolphe Brongniart.

Fig. 5 is a transverse section of the same part of the axis last described, magnified forty-five diameters.

Fig. 7 is a horizontal section of a single sporangium (magnified five diameters), showing its irregular pear-shaped form. The wall is composed of one line of transversely elongated cells; and the inside is a mass of microspores, many of which divide into three, some into four, and others into five sporules, all composed of a yellowish-brown hydrocarbon, resembling crude paraffine in appearance. In every respect this sporangium and its contents so closely resemble that of the cone described by the late Dr. Robert Brown that a description of one would nearly do for the other.

Fig. 8 is a vertical section (magnified four and a quarter diameters) of the pedicel and apex of a scale or bract, and of its underlying sporangium, that has an elongated oval form. The scale or bract is chiefly formed of cellular tissue, enclosing in its centre a bundle of vascular tissue, which is seen in the specimen to proceed from the axis and traverse the scale to its apex. The apex is broad, dilated at right angles to the pedicel, produced upwards into a triangular acute point, and downwards into a blunt lobe, as described by Dr. Hooker.[1] This sporangium was probably of an elongated oval form, and the notch shown in the lower part of this sporangium towards the axis is due to some disturbing cause, as it is not seen in the other sporangia of the cone.

Fig. 9 is a longitudinal section of the central axis of the cone, magnified forty-five diameters, showing the place of the pith (most probably composed of cellular tissue, but

[1] 'Mem. Geol. Surv.,' vol. ii, part 2, p. 450.

destroyed in the specimen), and the vascular cylinder of seven barred vessels on each side. Those next the pith being considerably larger than those on the outside.

Fig. 10 represents a longitudinal section of a portion of the wall of the sporangium, composed of a single line of transversely elongated cells, and a group of microspores, composed of crude paraffine, and dividing into three, four, and five sporules; but far more frequently into three; magnified fifty diameters.

This specimen in all its parts resembles Dr. R. Brown's Cone; and the only additional information that it affords is the structure of the central axis wanting in his specimen. The structure of the central axis in the Oldham specimen in all its parts, except the medulla of cellular tissue, is so closely identical with that of the stem of *Lepidodendron Harcourtii* (see Specimen No. 18, fig. 6, next described), that there can be little doubt as to its having been the fructification of that plant; and most probably Dr. Brown's specimen belonged to it also, as it exactly resembles my specimen, except that the central axis has been destroyed so far as structure is concerned. Now M. Brongniart is of opinion that Dr. Brown's *Triplosporites* is identical with his specimen. If this be so, the last-named genus has to be merged into *Lepidostrobus*,[1] if not to *Lepidodendron Harcourtii*.

Specimen No. 18, *Lepidodendron Harcourtii* (Plate VII, fig. 6).

Fig. 6 is the transverse section of the inner portion of a stem of *Lepidodendron Harcourtii*, magnified ten diameters, showing the central axis or pith composed of fine cellular tissue, surrounded by a zone of fine vascular tubes (of a hexagonal form, and having their sides barred with transverse striæ) and a sinuous boundary-line, of a dark colour, from which spring the vascular bundles that communicate with the leaves. The outer portion of the specimen clearly shows the band of lax cellular tissue, traversed by vascular bundles, and the outer radiating zone so usually found in stems of *Lepidodendron* exhibiting structure.

This specimen was presented to me by Mr. J. S. Dawes, F.G.S., and was found by him in the Dudley Coal-field. It is described and figured for the purpose of showing the identity of the structure of the stem of *Lepidodendron Harcourtii* with that of the axis of the Cone above described. If similarity of structure is of any value in proving the connection of organs of fructification with a stem, the Oldham Cone must be held to belong to *Lepidodendron Harcourtii*.

---

[1] As Dr. Brown himself and Mr. Carruthers also have shown; see 'Geol. Mag.,' vol. ii, p. 437.

Specimen No. 19, *Lepidodendron vasculare*.   Plate VIII, figs. 1—5, 7—9.

Fig. 1 is a fragment of a Cone, one and eight tenths of an inch in length, one and one tenth of an inch across its major, and one inch across its minor axis. This Cone is somewhat compressed out of its original cylindrical form, but not so much as is the Specimen No. 17. The fossil has lost the upper portions of the scales or bracts, in shelling out of its matrix, but it shows the rhomboidal scars of *Lepidostrobus*. They are not so broad as those of the last-described specimen. It also came from the Upper Foot Coal, near Oldham, and was found by Mr. John Butterworth, who has liberally allowed me to slice and describe it.

Fig. 2 is a transverse section of the specimen (natural size), showing the central axis of the Cone in a fair state of preservation; but the Sporangia connected with it are much disarranged, so that it is impossible to say how many there were in its original state.

Fig. 3 is a longitudinal section (natural size) of a portion of the Cone, showing the central axis and the scales or bracts, arranged in spiral order, and supporting Sporangia on each side of it.

Fig. 4 is a transverse section of the Cone, magnified two and a quarter diameters, showing the arrangement of the central axis, composed of hexagonal barred tubes, the smallest being towards the outside, where there is a dark line, nearly circular, and not so sinuous as in the last-described specimen. From this boundary-line spring bundles of barred vessels, that pass through the zone of cellular tissue, and communicate with the scales or bracts. These vessels traverse, in a highly inclined curve, a band of lax cellular tissue, which has for the most part been replaced by carbonate of lime. The outside of the axis exhibits elongated cells or utricles, arranged in radiating series, resembling those found in *Lepidodendron Harcourtii*. In the inner portion of the axis appear two circular bodies; one (in the upper part of the figure on the right hand side) having a white space, is the central axis of the plant, magnified in fig. 5; and the other (in the lower part on the left hand side) shows one of the vascular bundles which communicated with the scales or bracts.

Fig. 5 is also a transverse section of the inner portion of the central axis previously described, but magnified thirty-five diameters. The middle part is not very well preserved, but it sufficiently shows that the pith or medulla occupied by cellular tissue in *Lepidodendron Harcourtii* is here formed of barred tubes, like those in *Lepidodendron vasculare* and *Sigillaria vascularis*.

Fig. 7 is a horizontal section of a single Sporangium and a portion of a pedicel of a scale or bract, magnified five diameters. The former is so much disarranged that its original form cannot now be well recognised; and no trace of Microspores can be seen, the whole of the Sporangium having been changed into carbonate of lime. If any Microspores ever were in the Sporangium, they may have been shed or destroyed before the calcification of the Cone.

7

Fig. 8 is a vertical section of the pedicel and apex of a scale or bract (magnified four diameters) tolerably well preserved, together with part of the Sporangium which it supported, not so well preserved. The structure and shape of the bract is like that shown in Plate VII, fig. 8; but of the original form of the Sporangium there is not much evidence left. No trace of Microspores, or of paraffine, is now to be seen in it. The scale is chiefly formed of cellular tissue, enclosing in its centre a bundle of vascular tissue, which is shown in the specimen to proceed from the axis and to traverse the scale to its apex. This latter is broad, dilated at right angles to the pedicel, and produced upwards into a triangular acute point, and downwards into a blunt lobe, as described by Dr. Hooker in his *Lepidostrobus*.

Fig. 9 is the longitudinal section of the inner portion of the central axis of the Cone, magnified twenty-eight diameters.

In both specimens, No. 17 and No. 19, the chief value of the information they afford is in the structure of the axes of the two Cones. In the specimens treated of by Dr. Brown, Dr. Hooker, and MM. Brongniart and Schimper, the upper portions of the Cones (so far as the scales or bracts and Sporangia are concerned) have been fully described by the two former authors, and the whole of the Cone by the two latter; but there is no complete description of the structure of the central axis. My specimens appear to me to supply to a great extent that deficiency. We have seen how the whole of the specimen " No. 17 " not only agrees with that of Dr. Brown, but that its axis is identical in structure with that of *Lepidodendron Harcourtii*. M. Brongniart appears to think that Dr. Brown's specimen was merely the upper part of a Cone similar to his, called by Prof. Schimper *Lepidostrobus Dabadianus*. If this should be proved to be the case, both my " No. 17 " and Dr. Brown's specimen may probably prove to be the upper portion of a Cone with Sporangia containing Microspores, having its lower portion composed of Sporangia containing Macrospores, and thus prove that all these three cones are the fructification of *Lepidodendron*, and, not unlikely, that of *L. Harcourtii*.

No. 19 bears considerable resemblance in structure, so far as its central axis is concerned, to the stem of *Lepidodendron vasculare* (see below); and therefore it is here referred to that plant, but not without doubt. One thing is certain that, although externally it is like No. 17, it is quite a different Cone so far as its internal structure is concerned. This specimen also, probably owing to its advanced stage of growth, had shed its Spores before it was mineralized.

Specimen No. 20; *Lepidodendron vasculare*. Plate VIII, fig. 6.

Fig. 6 is a transverse section of the inner portion of a stem of *Lepidodendron vasculare*, magnified twelve diameters, showing the centre, composed of hexagonal tubes of barred vessels of different sizes, more regular and of less diameter towards the outside. There is also an outer zone of tissue, with interruptions in it.

This is a representation of the specimen described and figured (p. 110, Plate VI, figs. 2 and 3) in my paper " On some Fossil Plants, showing Structure, from the Lower Coal-Measures of Lancashire."[1] It is here reproduced for the purpose of showing the structural similarity of the axis of the Cone now under consideration with that of *Lepidodendron vasculare*, and its difference from *Lepidodendron Harcourtii* (see above, page 48).

§ 2.—SPECIMENS Nos. 21 and 22 ; *Lepidostrobus Russellianus*, sp. nov. Pl. IX, figs. 1, 1 *a*, 2, 2 *a*.

Specimen No. 21, *Lepidostrobus Russellianus* (Plate IX, fig. 1), natural size, is a compressed imperfect Cone, six inches in length, by eight tenths of an inch in breadth, having a central column one tenth of an inch in diameter. The upper portion of the fossil is not preserved, so its form is unknown; but the lower portion, for about two inches, shows the scales or bracts of the Cone, springing nearly at right angles to the column, and arranged in spiral order, together with numerous disc-shaped bodies, about one thirty-second of an inch in diameter, having their insides smooth and coriaceous, and their outsides granular, but showing no clear evidence of a triradiate ridge, although there is some sign of an elevation on some of the discs.[2] The bracts on the upper part of the Cone do not show such bodies.

The matrix in which the fossil is imbedded is a Black Band Ironstone; but the disc-shaped bodies are chiefly composed of granular bisulphide of iron, coating the coriaceous layer, of a yellowish colour, on their outsides; whilst their insides are full of a granular substance of a bright-yellow colour, also resembling bisulphide of iron. The column, scales, and Sporangia are all converted into coal, and as yet have afforded no evidence of their former structure.

This and the four next described specimens are from the Coal-measures, near Airdrie, Scotland, and were found by Mr. James Russell, of Chapelhall.

Fig. 1 *a* (magnified five diameters), represents a portion of the column of the Cone and two Sporangia, each containing fifteen of the disc-shaped bodies, in seven pairs and one at the end. The scale or bract goes out nearly at right angles from the column to the end of the Sporangium, when it turns upwards nearly parallel to the stem. Both the scale and Sporangium have been converted into a mass of coal. The discs are both concave and convex; and some of them appear as if they had been separated into two by the splitting of the specimen.

[1] ' Quart. Journ. Geol. Soc.' for May, 1862, vol. xviii.

[2] In this, as well as in the other specimens of macrospores hereinafter described, no conclusive evidence of a triradiate ridge has been observed.

Their arrangement appears to me to show that they were contained in a Sporangium, and were Macrospores, rather than separate Sporangia, attached to the surface of the scales, as Mr. Carruthers described was the case in his specimen;[1] and hereinafter they will be termed Macrospores.

This specimen also shows the Macrospores only on the lower, and not on the upper, part of the Cone. The outside of the scales is not very well shown; but their quincuncial arrangement and the form of the scar, where they were attached to the column, are both very clearly shown, and cannot be distinguished from such parts in *Lepidostrobus* or the leaf-scars of *Lepidodendron*. It is named *Lepidostrobus Russellianus* from Mr. Russell, who discovered the specimen.

Specimen No. 22, *Lepidostrobus Russellianus.*   Plate IX, figs. 2 and 2 *a.*

Fig. 2, of natural size, is another compressed, imperfect Cone of *Lepidostrobus Russellianus*, four inches long, nine tenths of an inch in breadth, and having a column one tenth of an inch across. The upper part is wanting; but what of the lower portion now remains shows Sporangia, of a somewhat oval form, springing nearly at right angles from the column, and full of Macrospores, only one thirty-second of an inch in diameter but similar in all other respects to those described in the last specimen, No. 21; and like it, the base and apex of the Scale or Bract and the Sporangium-wall are not well shown, all these parts being converted into bright coal, so that none of them can be clearly distinguished. The scars on the column of the Cone shows that the Scales were arranged spirally around the axis.

Fig. 2 *a* (magnified five diameters) shows a portion of the column of the Cone, with traces of the Bract-scars, and a single Sporangium, containing fourteen Macrospores, arranged in a double series, like those in Specimen No. 21, except that the odd one at the end is here wanting. The Macrospores have their outsides formed of a yellowish paraffine, covered with granular bisulphide of iron. Their insides contain bright granules of bisulphide of iron, which at first sight might be mistaken for Sporules.

As this Cone is not to be distinguished from No. 21, it has also been called *Lepidostrobus Russellianus.* It is also from a Blackband Ironstone near Airdrie.

§ 3. Specimen No. 23, *Lepidostrobus* (?) *dubius*, sp. nov.   Plate IX, figs. 3, 3 *a.*

Specimen No. 23, Plate IX, fig. 3, is another imperfect, compressed Cone, from the Blackband Ironstone, near Airdrie, four inches in length, six tenths of an inch in breadth, with a column one twelfth of an inch wide. All the Cone exposed shows

[1] 'Geological Magazine,' vol. ii (No. XVI), p. 434.

Scales or Bracts springing almost at right angles to the column, and supporting Sporangia, full of Macrospores, one twenty-fifth of an inch in diameter, very similar in their characters and state of preservation to those contained in the two specimens described above. The apices of the Scales are both longer and stronger than those of No. 21 and No. 22. The column is not so well exposed, and the lower portion is covered up in the matrix of Blackband Ironstone, so that the connection of the Cone with the associated stem, striated, knotted, and jointed, resembling a small Calamite, cannot be traced, a quarter of an inch intervening. The occurrence of this stem may be accidental, having no connection with the Cone, but the column of the latter, if projected forwards, would run to the joint of the stem ; and as there are two specimens, No. 27 and No. 30, connected with somewhat similar stems and containing Macrospores, to be hereinafter described, it is possible that this Cone may have belonged to a different plant than those of No. 21 and No. 22. For this reason it has been designated *Lepidostrobus* (?) *dubius*.

Fig. 3 *a* (magnified five diameters) represents a portion of the column and two Scales supporting two Sporangia, each containing three large Macrospores, with several smaller ones; the upper Sporangium has eight smaller ones, in two rows, and one by itself at the end ; whilst the lower Sporangium contains eleven of the smaller Macrospores, in five pairs, and one at the end. In all other respects the Sporangium and its Macrospores, in their present state of preservation and contents, cannot be distinguished from those previously described. The presence of the striated, jointed, and knotted stem, and the different sizes and arrangement of the Macrospores, may be differences indicating that this specimen is more allied to No. 27 and No. 30, hereinafter described, than to No. 21 and No. 22. As there is no evidence to show that the Scales were spirally arranged round the column, it is possible that they may have had a verticillate arrangement, as in specimens to be hereinafter described.

§ 4. Specimen No. 24 ; *Lepidostrobus tenuis*, sp. nov. Pl. IX, figs. 4, 4*a*.

Specimen No. 24 (Pl. IX, fig. 4, natural size) represents another imperfect compressed Cone, two and two tenths of an inch long, six tenths of an inch broad, and having a column about one fortieth of an inch thick. Both the upper and lower portions of the Cone are wanting ; but the whole of the specimen shows Scales or Bracts, supporting elongate-oval Sporangia, containing Macrospores, one twenty-fifth of an inch in diameter, in a similar condition to the specimens previously described. The only remarkable feature in this specimen is the great delicacy of the Scales and Column, which are much less in size than in any of the other specimens. The apex of the scale also is more divergent, and is not so parallel to the column as in No. 21.

Fig. 4a (magnified 5 diameters) shows two Scales and two Sporangia, each containing Macrospores, all nearly of the same size; fourteen are seen in the upper Sporangium, and fifteen in the lower one. The apex of the Scale is stronger, and diverges more from the vertical direction than most of the specimens previously described.

This *Lepidostrobus* may be of the same species as No. 21 and No. 22, but for the present it is probably better to distinguish it as a provisional species by the name of *tenuis*.

§ 5. SPECIMEN No. 25 ; *Lepidostrobus levidensis*. Pl. X, figs. 1, 1a, 1b.

Specimen No. 25 (Pl. X, fig. 1, natural size) represents a beautiful Cone in a compressed state, and nearly perfect, six and a half inches in length and eight tenths of an inch in breadth; its central column or axis measuring one tenth of an inch across. This is longer, and tapers more than any of the specimens previously described. Excepting a small portion, it shows the apex. The base of the Cone is broken off, so we cannot now see how far it may have extended; but for the space of an inch the lowest part of the Cone shows Scales or Bracts arranged in spiral order around the column, springing out at right angles to it, and supporting elongate oval Sporangia, full of Macrospores, one twentieth of an inch in diameter. The upper portion of the Cone is for the most part covered with rhomboidal scales; but on the right hand side, near the top, are exposed some oval Sporangia, containing very small spheroidal bodies, converted into a substance like paraffine, and resembling Microspores. Thus we have, in this case, a Cone, with the form and external characters of a *Lepidostrobus*, possessing two kinds of Sporangia, namely the lower series inclosing Macrospores similar to those described in the previous specimens, and the higher set containing Microspores resembling those in *Lepidostrobus Dabadianus*, Schimper.

Fig. 1a (magnified 2½ diameters) represents a portion of the Column of the Cone, showing the Scars of the Bracts, arranged in quincuncial order, and four Bracts on each side of the Column, at right angles to it, and supporting long-oval Sporangia, containing from fourteen to sixteen Macrospores in each. The Macrospores are chiefly composed of granular bisulphide of iron on a coriaceous covering of a yellowish colour. Their insides consist also of grains, resembling bisulphide of iron. The Column, Bracts, and Sporangia are converted into coal, and as yet have afforded no evidence of their former structure.

Fig. 1b (magnified 5 diameters) represents a Bract, supporting an irregularly oval Sporangium, full of small spheroidal bodies, converted into a yellowish substance like paraffine, and resembling Microspores. Of course these Microspores are not so clear and distinct as those seen in Specimen No. 17, by the aid of transmittent light; but they are of the same substance, and much like those bodies when seen by reflected light. They appear to be about the same size as those found in *Lepidodendron Harcourtii*.

§ 6. SPECIMEN No. 26 ; *Lepidostrobus Hibbertianus*, sp. nov. Pl. X, figs. 2, 2*a*, 2*b*.

Specimen No. 26 (Pl. X, fig. 2, natural size) represents another compressed Cone, nearly perfect, six inches in length, and nine tenths of an inch in breadth, with an axis one tenth of an inch across. A portion of the apex is wanting, but the greater part of the base is left, so that this Cone is in a more perfect condition than those previously described. For the greater part it exhibits a surface of broad, rhomboidal, and imbricated Scales (ends of Bracts), arranged spirally ; but for about an inch of the lower portion stout Bracts are seen springing from the column at right angles, and bearing long-oval Sporangia, full of Macrospores, one twenty-eighth of an inch in diameter.

In its external character this specimen is not to be distinguished from *Lepidostrobus ornatus*, and most collectors would class it in that species. The only remarkable features it possesses are the Sporangia containing Macrospores.

This Cone is imbedded in Burdiehouse Limestone, from near Edinburgh, and not in Blackband Ironstone, as the five preceding specimens are ; but the Column and Bracts are converted into coal, and the coriaceous covering of the Macrospores is of a yellowish matter, like crude paraffine, as in those other specimens. This fossil, from the Collection of the late Dr. Hibbert, of Edinburgh, was purchased by me at the sale of his museum.

Fig. 2*a* (magnified 5 diameters) represents a portion of the central Column and two well-defined Scales or Bracts, running at right angles from the Column, afterwards turning upwards nearly parallel to it, and supporting two long-oval Sporangia, full of Macrospores. In each Sporangium there are seen sixteen Macrospores, seven in the upper and eight in the lower series, with a terminal one.

Fig. 2 *b* (magnified five diameters) represents four of the rhomboidal scales on the upper part of the specimen.

This Cone[1] is named after that well-known geologist, Dr. Hibbert, of Edinburgh, in whose collection it was found.

§ 7. SPECIMEN No. 27 ; *Lepidostrobus* (?) *ambiguus*. Pl. XI, figs. 1, 1*a*, 1*b*.

Specimen No. 27, Plate XI, fig. 1 (natural size), represents the stem and the lower part of a compressed Cone, the upper portion of which is wanting. Exclusive of the Stem it is one and one tenth of an inch in length, and four tenths of an inch in breadth. The leaves connected with the Stem, and the remaining Bracts of the Cones appear to have been

---

[1] It is desirable that the upper portion of this Cone should be carefully ground down, in order to ascertain whether or not it contains Sporangia, full of Microspores, similar to those found in the specimen last described. It appears to me that this could be done with some little trouble.

arranged in whorls, and not spirally, as the Bracts are in most of the specimens previously described. In the upper part the Bracts go off from the column nearly at right angles, supporting oval Sporangia; whilst in the lower part the Bracts are inclined downwards at a considerable angle. The Sporangia all contained Macrospores, about one twenty-fifth of an inch in diameter; but, owing to compression, their characters are not well shown.

This interesting specimen, consisting probably of not even all the lower half of the original cone, was found in the Carboniferous Ash-beds at **Laggan Bay**, in the Isle of Arran, by Mr. Russell.

Fig. 1*a* represents the lower part of the Cone and the upper portion of the stem, to which it is attached, magnified five diameters. The Leaves connected with the stem are not very well defined; but they appear to have had a verticillate arrangement. The Sporangia in the lower part of the Cone were pear-shaped, and inclined downwards; and, from signs of Macrospores in the specimen, they appear to have contained those bodies, which are not well shown.

Fig. 1*b* (magnified five diameters) represents the crushed upper portion of the specimen; and in it are seen some indistinct Macrospores.

This imperfect specimen is here figured and described for the purpose of showing that some of these Cones have their Leaves and Scales arranged verticillately. It bears some resemblance to No. 23 previously described, and to No. 30 hereinafter described (Plate XII, fig. 1). It is named provisionally *Lepidostrobus ambiguus*.

§ 8. SPECIMEN No. 28; *Lepidostrobus Wuenschianus*, sp. nov. Pl. XI, figs. 2, 2*a*, 2*b*, 2*c*.

Specimen No. 28, Pl. XI, fig. 2 (natural size), represents a slender cone, one and eight tenths of an inch in length, and four tenths of an inch in breadth, found in the ash-beds at Laggan Bay by Mr. Russell. A portion of the Cone next the Stem is unfortunately wanting, but the specimen, on the whole, is in a more perfect condition than any of those previously described. The Sporangia are of an irregular oval form, and are supported by Scales or Bracts, which, in the middle of the specimen, spring from the Column at right angles, show a spiral arrangement, and have their apices pointing upwards nearly parallel to it. The Sporangia contain Spores of two kinds: the upper ones being filled with a granular matter, composed of small spheroidal Microspores; whilst the lower Sporangia, on each side of the column for the last four, so far as exposed, had Macrospores one twentieth of an inch in diameter, with granulated outsides. In each of the latter Sporangia there appears, from the evidence afforded by the specimen, to have been three Macrospores.

Fig. 2*a* (magnified five diameters) represents the upper part of the Cone, showing the apex and six Scales on each side of the Column, supporting Sporangia filled with fine granular matter.

Fig. 2*b* (magnified five diameters) represents the lower part of the Cone next the Stem, and exhibits evidence of four Scales, on each side of the Column, supporting Sporangia, containing Macrospores.

Fig. 2*c* (magnified ten diameters) represents a single oblong-oval Sporangium, full of spheroidal Microspores, from the upper part of the Cone. The Microspores are very small, and cannot be separated for measurement with any degree of certainty, owing to their having been converted into a bright substance resembling iron-pyrites.

Although this Cone is the least of any of the specimens described, its Macrospores are some of the largest. It has been named after Mr. Wünsch, who first discovered the true nature and position of the ash-beds in the Carboniferous formation at Laggan Bay.[1]

§ 9. SPECIMEN No. 29 ; *Lepidostrobus latus*, sp. nov. Pl. XI, figs. 3, 3*a*, 3*b*, 3*c*.

Specimen No. 29, Pl. XI, fig. 3 (natural size), represents a stout Cone, of an oval form, nearly two inches in length by seven tenths of an inch in breadth in its middle, from the ash-beds of Laggan Bay ; also found by Mr. Russell. This is by far the most robust Cone that has come under my notice. It was covered by strong Scales ; each Bract having a thick median rib, and taking nearly a vertical direction, as shown in the upper portion of the specimen. In its middle and lower portions the specimen has lost its Bracts, and affords evidence, although not so well marked as in the specimen last described, of two kinds of Sporangia ; the upper ones exposed being pear-shaped, and containing apparently Microspores ; and the lower ones which slope downwards have irregularly discoidal bodies like Macrospores.

Fig. 3 *a* (magnified five diameters) shows the upper portion of the Cone and its stout, ribbed Scales.

Fig. 3 *b* (magnified five diameters) shows the lower part of the Cone with its Sporangia, and Scales inclining downwards, and some bodies resembling Macrospores.

Fig. 3 *c* (magnified fifteen diameters) shows a pear-shaped Sporangium, from the upper part of the Cone, filled with small granular bodies resembling Microspores.

The three specimens last described are, as previously stated, from the trap-ash of Laggan Bay, the same deposit as that from which the beautiful stems of *Sigillaria*, *Lepidodendron*, *Halonia*, *Antholithes*, and other Coal Plants discovered by Mr. Wünsch, of Glasgow, were obtained. The ash enveloping the stems is sometimes nearly as hard as greenstone, and at other times quite soft and pulverulent. It is of a greyish colour, contains small bright pieces of iron-pyrites, and is intersected by narrow veins of common and fibrous carbonate of lime : the mass effervesces when treated with hydrochloric acid. By the kindness of Mr. Wünsch, I am enabled to give the chemical composition of this ash.

[1] 'Transact. Geol. Society of Glasgow,' vol. ii, p. 97.

" It consists of carbonates, soluble and insoluble silicates, quartz, and iron-pyrites.   The carbonates are removed by dilute acetic acid, and consist of carbonate of lime, with a very little of carbonate of iron, and carbonate of magnesia.   The insoluble silicates consist principally of the silicate of alumina, with a little lime, magnesia, potash, and soda.

" Composition of the 'Ash,' after drying it at 212° Fah. :

Silicic acid	13·20
Ferrous oxide	18·26
Manganese	·78
Alumina	8·13
Lime	13·47
Magnesia	5·06
Carbonic acid	8·40
Insoluble silicates	28·76
Bilsulphide of iron	·70
Water	3·23
	99·99

(Left brace label: Soluble silicates and carbonates.)

Specific gravity 2·790.

" Analysis of a stem of *Lepidodendron* from the Ash, previously dried at 212° Fah.

Carbonate of Lime	89·16
Carbonate of Magnesia	1·26
Carbonate of Iron	1·06
Carbonate of Manganese	2·22
Ferric Oxide	1·39
Insoluble matter	2·24
Carbon	2·44
	99·77

Specific gravity 2·611."

The analysis of this fossil plant shows apparently that the decomposing wood had the power of attracting the carbonate of lime from the surrounding matrix in which it was imbedded.

Both the form and the structure of the fossil plants in the ash have been beautifully preserved.   They remind us much of the plants found in the Coal-measures of Lancashire, so far as their genera and species are concerned.   From the appearances which they now present, we see they were growing in water, on the spots where they are now found, when fine ash, erupted from a neighbouring volcano, quietly and gradually enveloped them in the matrix in which they occur.   We find the fragile leaves of the most delicate *Sphenopteris*, and the fine-pointed leaves of *Lepidodendron*, just as they grew, without the slightest fracture or disarrangement.   Nearly all the stems appear to have suffered little from compression or disturbance in their parts.   (See notices of this trappean ash interstratified with the Lower Carboniferous series of Arran, in the ' Geol. Mag.,' vol. ii, 1865, p. 474 ; vol. iv, 1867, p. 551 ; ' Trans. Geol. Soc.,' Glasgow, vol. ii, part 2, 1866, p. 97, &c.).

§ 10. SPECIMEN No. 30 ; *Bowmanites Cambrensis*, gen. et sp. nov.   Plate XII, figs. 1, 1 *a*, 1 *b*, 1 *c*, 2, 3.

Pl. XII, fig. 1 (natural size), and fig. 1 *a* (magnified two diameters), represent a branch and part of the lower portion of the Cone of a singular plant, found many years since by the late Mr. John Eddowes Bowman, F.G.S., in a nodule of clay-ironstone at the Varteg Iron-works, near Pontypool, South Wales.   For these specimens, my thanks are due to his son, the late Professor Eddowes Bowman, who liberally presented them to me.   The branch proceeded from the stem at the hole seen in the lower part of the specimen (fig. 1), and consisted of a slight, ribbed and furrowed, cylindrical stem, parted at regular intervals by joints and knots, giving rise to verticillate leaves ; it was terminated by a long Cone, cylindrical in the middle, and tapering at its extremities.   The whole of the substance of the Stem and Cone was replaced by a white powder, so that only a mould of the fossil has been left in the matrix of clay-ironstone, with the exception of dark-coloured and granulated discs, about one twenty-second of an inch in diameter, which have been left on the sides of the mould, in the exact position which the corresponding spots occupied in the original plant.

The leaves of the plant came out in whorls of sixteen, at each knot, and were of a subulate form, and strongly ribbed in the middle.   They resemble those of *Asterophyllites*.

The form of the stem and branch remind us of the *Bechera grandis* of Lindley and Hutton ; but the shape and characters of the Cone and its contents differ very considerably from those of that plant.

Fig. 1 *a* (magnified two diameters) represents a portion of the stem and of two whorls of leaves, as they now appear in the specimen.

Fig. 1 *b* (magnified two diameters) represents a cast, taken in gutta-percha, of a portion of the stem, showing its ribbed and furrowed surfaces, and the joints and knots, whence proceeded the leaves, apparently sixteen in each whorl.

Fig. 1 *c* (magnified two diameters) represents a cast, in gutta-percha, of the branch to which the Cone was attached ; its ribs, furrows, and knots, and the origin of the leaves are very distinct.

As previously stated, the two specimens of this plant belonged to the late Mr. J. E. Bowman.   More than thirty years since, when my late friend first showed them to me, the whole consisted of six or seven pieces, comprising apparently the entire original nodule.[1]   At that time he had made an enlarged drawing of the whole of the specimen,

[1] Mr. Wm. Bowman, F.R.S., has kindly allowed me to search his late father's cabinet, but the missing parts of the fossil have not yet been found.

restoring the plant, according to his idea of its form and character, from the parts then in his possession. Fig. 2 (natural size) represents the restored Stem, Branch, Leaves, and Cone, as it appeared to Mr. Bowman. This is copied from his original drawing.

Fig. 3 (magnified six times) represents, according to Mr. Bowman's view, and from his original drawing, two of the lower Scales, each supporting five Macrospores, but not giving evidence of the walls of any Sporangium that enclosed them.

In fig. 1 *a*, where the two whorls of Leaves are seen, they appear to be distinct and separate, springing from the rounded knobs at the joints, just as they appear on the stem and branch, figs. 1 *b* and 1 *c*. Now, in Mr. Bowman's restored drawing (fig. 2), he makes the leaves united at their base, and springing from a kind of sheath that embraces the stem to which they were attached. No doubt Mr. Bowman, who was a skilled botanist, and had better information than we possess, must have been better qualified to speak with certainty on this point than, from the two fragments in our possession, we are now.

Whatever may be the true characters of the leaves of this plant, it undoubtedly furnishes us with evidence of the former existence of a stem, with verticillate leaves, possessing a Cone with Macrospores in its lower part ; and thus it induces us to believe that Specimens No. 23 and No. 27, hereinbefore described, may be more nearly allied to this plant than to the genus *Lepidostrobus* with which they have been, in doubt, provisionally classed.

At first, from the characters of the stem and leaves, it occurred to me to place this Cone in the genus *Calamostachys* of Schimper ; but the Macrospores in it are so different to the spores of *Calamostachys* that it is probably better to establish a new genus. It is, therefore, here called *Bowmanites*, after its discoverer, Mr. Bowman, and *Cambrensis* from the country where it was found.

V. CONCLUDING REMARKS.

This Monograph, no doubt the reader will have perceived, was intended to be of a descriptive character rather than an attempt to trace the analogy of those plants, the remains of which have formed our valuable beds of coal, with living vegetables. My endeavours have been to collect materials and give them to the public for botanists to work upon. The subject is surrounded with difficulties ; and, although it has been my good fortune to meet with many specimens in a fair state of preservation, the specimen, as a rule, when the internal structure is well preserved, is in a fragmentary condition, and when several parts of a plant are found connected together we are not favoured with structure, as is the case of the beautiful fossil plant last described.

When we consider how common a fossil a *Lepidostrobus* is, met with in abundance in all our Coal-measures, and described in nearly every work on Carboniferous fossils, it is very remarkable how few of the specimens afford us much evidence of the true nature of their organs of fructification.

Until the description of *Lepidostrobus Dabadianus* was given by M. Brongniart and Professor Schimper, we were ignorant of a fossil Cone with Sporangia full of Microspores in its upper part, and Macrospores in its lower part; and even this valuable specimen was found in drifted deposits, so we cannot be quite certain as to its having originally come from the Coal-measures. True it is that both Dr. Robert Brown and Dr. Hooker adduced evidence of Cones with Sporangia containing Microspores; but those distinguished authors never stated that such Cones might also have contained Macrospores in their lower portions.

In addition to *Lepidostrobus Dabadianus* two more Cones, namely, *Lepidostrobus levidensis* and *Lepidostrobus Wuenschianus*, and probably a third, *L. latus* (all from undoubted Carboniferous Strata) have to be added to the list of Cones with both Microspores and Macrospores; whilst seven other Cones, also from the Coal-formations, have been described, which, so far as they can be examined, afford evidence of Macrospores alone.

The Cone first described in this Monograph (No. 17) might, as far as its external characters are concerned, be taken for *Lepidostrobus ornatus;* and it contains Microspores in a most beautiful state of preservation, not to be distinguished from those found in Dr. Brown's Cone and in *Lepidostrobus Dabadianus.* On the other hand, in my Cone, No. 26 (*Lepidostrobus Hibbertianus*), which would also pass for a good example of *L. ornatus*, we cannot see the Sporangia in its upper portion, owing to the Scales, but where the interior of the lower part is exposed we meet with Sporangia containing Macrospores, like those in *L. Dabadianus.* This leads us to conclude that similar Cones, well preserved, on being subjected to careful examination, will afford the two kinds of spores, in the upper and lower portions respectively.

In nearly the smallest Cone described (No. 28, *L. Wuenschianus*) the largest Macrospores were found; thus showing that the size of the Cone had nothing to do with producing a large Macrospore.

The Cones No. 23 and No. 27 have been classed under *Lepidostrobus* with considerable doubt, as there is not sufficient evidence, especially as regards No. 23, to place them with *Calamostachys.* They are evidently fragments, and the lower portions only of two Cones; they may, therefore, have had Microspores in their upper parts. The last-described Cone (No. 30) has been referred to a new genus (*Bowmanites*), as it differs from *Calamostachys* in its organs of fructification. An important feature in these three Cones is that they appear to have had a verticillate arrangement of Scales, with whorled Leaves on their Stems, and that in their lower portions they have yielded Macrospores.

The Macrospores in all the specimens appear to be of large size, when compared with those found in *Lepidostrobus Dabadianus*; but it must be borne in mind that the former have been much compressed, and even flattened out, while the latter retain their original form. This will, to a certain extent, account for the apparent difference in size. No doubt there might be various sizes of both Microspores and Macrospores in the plants then existing; and, indeed, we could scarcely expect to find them all of one size. In the flora of the Carboniferous epoch, Cones having Sporangia with two kinds of spores appear to have constituted a more marked character of the period than has been hitherto supposed.

In the calcareous nodules found in the Lower Brooksbottom Seam of Coal, as well as that of the Upper Foot Coal, plenty of detached Macrospores are to be met with, and in a few instances in or near a Sporangium. There are also numerous traces of Microspores to be found in the same nodules, when the slices have been ground down fine enough; but this is not very easily done, for the paraffine-like matter of which the spores are composed is apt to tear away in the operation of grinding. With all this allowance, however, it must be admitted that in neither of these seams of coal are Macrospores to be met with in anything like the quantity in which they are found in the "splint" and "brown cannel" coals of Scotland.

PLATE VII.

Lepidodendron Harcourtii, L. and H.

Fig. 1 (No. 17). The imperfect Cone from the Upper Foot seam of Coal, near Oldham, as it appeared before it was sliced. Natural size.

Fig. 2. A transverse section. Natural size.

Fig. 3. A longitudinal section. Natural size.

Fig. 4. A transverse section (fig. 2). Magnified 2 diameters.

Fig. 5. A transverse section of the pith and vascular cylinder. **Magnified 45** diameters.

Fig. 6 (No. 18). A transverse section of the pith and vascular cylinder of *Lepidodendron Harcourtii*, from the Dudley Coal-field. Magnified 10 diameters.

Fig. 7 (No. 17). A transverse section of a single Sporangium, full of **Microspores.** Magnified 5 diameters.

Fig. 8. A longitudinal section of a single Sporangium, with a Scale or Bract. **Magnified** $4\frac{1}{4}$ diameters.

Fig. 9. A longitudinal section of the pith and vascular cylinder. Magnified 45 diameters.

Fig. 10. A group of Microspores (converted into a substance resembling paraffine), and a portion of the wall of the Sporangium containing them. Magnified 50 diameters.

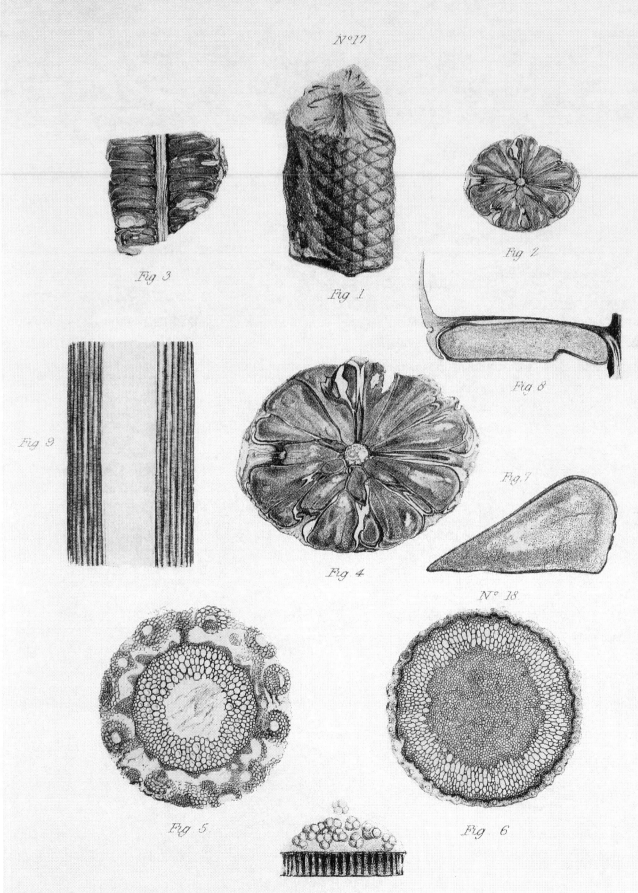

Plate VII.

N°17

Fig 3

Fig 1

Fig 2

Fig 8

Fig 9

Fig 4

Fig.7

N° 18

Fig 5

Fig. 6

Fig. 10

J. N. Fitch, del. et lith.

R. Fitch, imp.

PLATE VIII.

Lepidodendron vasculare, Binney.

Fig. 1 (No. 19).   The imperfect Cone from the Upper Foot seam of Coal, Oldham, as it appeared before it was sliced.   Natural size.

Fig. 2.   A transverse section.   Natural size.

Fig. 3.   A longitudinal section.   Natural size.

Fig. 4.   A transverse section (fig. 2).   Magnified 2¼ diameters.

Fig. 5.   A transverse section of the pith and vascular cylinder.   Magnified 35 diameters.

Fig. 6 (No. 20).   A transverse section of *Lepidodendron vasculare,* from the " Bullion Mine" Spa, Clough, near Burnley, showing the central axis and medullary sheath. Magnified 12 diameters.

Fig. 7 (No. 19).   A transverse section of a single Sporangium.   Magnified 4 diameters.

Fig. 8.   A longitudinal section of a single Scale or Bract and its Sporangium. Magnified 4 diameters.

Fig. 9.   A longitudinal section of the pith and vascular cylinder.   Magnified 28 diameters.

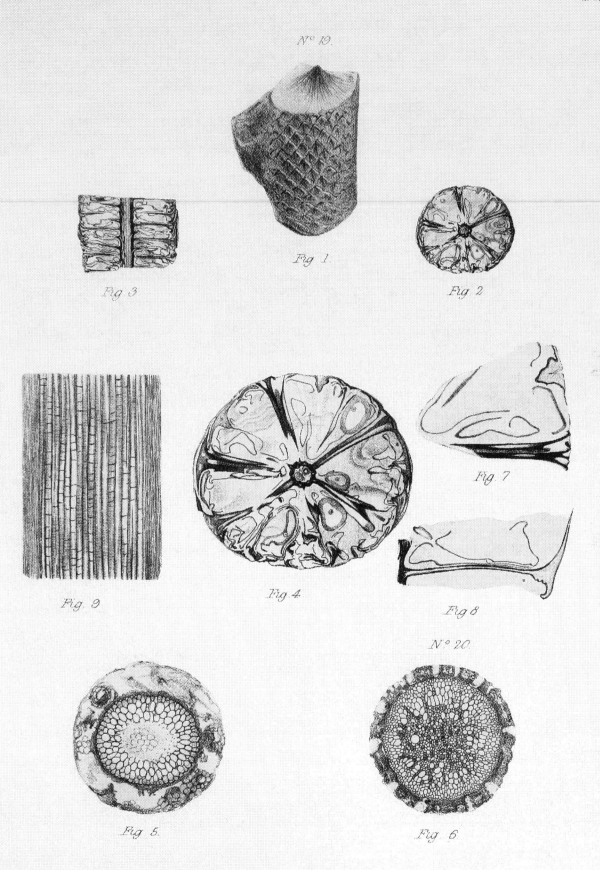

Plate VIII.

N° 19.

Fig. 3

Fig. 1.

Fig. 2

Fig. 9

Fig. 4.

Fig. 7

Fig. 8

N° 20.

Fig. 5.

Fig. 6.

J.N. Fitch del. et lith.

R. Etch. imp

PLATE IX.

Lepidostrobus Russellianus, Binney.

Fig. 1 (No. 21). A Cone, not perfect, from the Blackband Ironstone, near Airdrie, Scotland. Natural size.

Fig. 1*a*. A portion of the column of the Cone, and two Scales, supporting two Sporangia, full of Macrospores. Magnified 5 diameters.

Fig. 2 (No. 22). An imperfect Cone from the Blackband Ironstone, near Airdrie. Natural size.

Fig. 2*a*. A portion of the Column, and a Sporangium, full of Macrospores. Magnified 5 diameters.

Lepidostrobus ? dubius, Binney.

Fig. 3 (No. 23). The imperfect Cone from the Blackband Ironstone, near Airdrie. Natural size.

Fig. 3*a*. A portion of the Column, and two Scales, supporting two Sporangia, full of Macrospores. Magnified 5 diameters.

Lepidostrobus tenuis, Binney.

Fig. 4 (No. 24). A fragment of a Cone from the Blackband Ironstone, near Airdrie. Natural size.

Fig. 4*a*. Two Scales, and two Sporangia, full of Macrospores. Magnified 5 diameters.

Plate IX

N.º 23

N.º 21   4ª

N.º 24   3ª

Fig 3

Fig 4

Fig 1

N.º 22

Fig 2

2ª

1ª

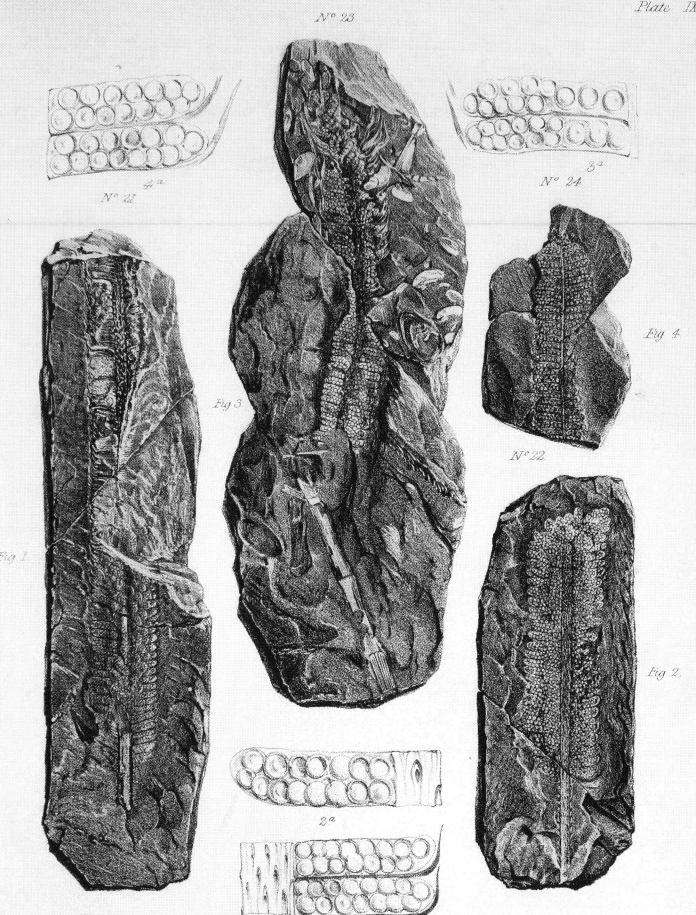

PLATE X.

Lepidostrobus levidensis, Binney.

Fig. 1 (No. 25). A nearly perfect Cone from the Blackband Ironstone, near Airdrie. Natural size.

Fig. 1*a.* A portion of the lower part of the Column, and eight Scales, supporting eight Sporangia, containing Macrospores. Magnified $2\frac{1}{2}$ diameters.

Fig. 1*b.* A Scale from the upper portion, and a Sporangium, containing Microspores. Magnified 15 diameters.

Lepidostrobus Hibbertianus, Binney.

Fig. 2 (No. 26). An almost perfect Cone, from the Burdiehouse Limestone (Lower Carboniferous), near Edinburgh. Natural size.

Fig. 2*a.* A portion of the Column of the lower portion, and two Scales, supporting Sporangia, full of Macrospores. Magnified 5 diameters.

Fig. 2*b.* Four of the rhomboidal Scales, or ends of Bracts, from the upper part of the Cone. Magnified 5 diameters.

Plate X

N° 25.

1 b

2 b

N° 26

Fig.1

Fig.2

2 a

1 a

N.Fitch.del et hth.

R. Fitch.imp.

PLATE XI.

Lepidostrobus ambiguus, Binney.

Fig. 1 (No. 27). The lower portion of a Cone, from the trappean ash of Laggan Bay, in the Isle of Arran. Natural size.

Fig. 1a. A portion of the Stem, and the base of the Cone. Magnified 5 diameters.

Fig. 1b. The upper part of the specimen. Magnified 5 diameters.

Lepidostrobus Wuenschianus, Binney.

Fig. 2 (No. 28). A Cone, from the trap-ash, Laggan Bay. Natural size.

Fig. 2a. The upper portion of the Cone, showing the Column, Scales, and Sporangia. Magnified 5 diameters.

Fig. 2b. Part of the lower portion of the Cone, showing the Column, Bracts, and Sporangia, with Macrospores. Magnified 5 diameters.

Fig. 2c. A Scale, from the upper part, with a Sporangium, full of Microspores. Magnified 10 diameters.

Lepidostrobus latus, Binney.

Fig. 3 (No. 29). A Cone, from the trap-ash, Laggan Bay. Natural size.

Fig. 3a. The upper portion of the Cone. Magnified 5 diameters.

Fig. 3b. The lower portion of the Cone. Magnified 5 diameters.

Fig. 3c. A Sporangium, from the higher part of the Cone, containing Microspores. Magnified 10 diameters.

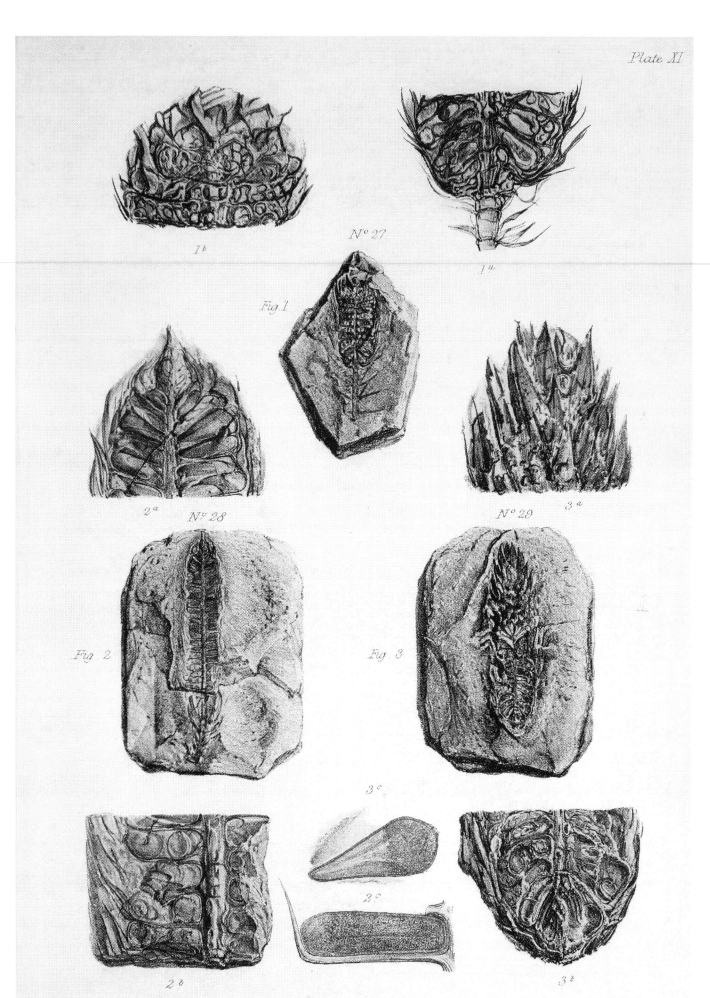

Plate XI

1ᵇ

Nº 27.

1ᵃ

Fig 1

2ᵃ   Nº 28

Nº 29.   3ᵃ

Fig 2

Fig 3

3ᶜ

2ᶜ

2ᵇ

3ᵇ

J. N. Fitch, del. et lith.

R. Fitch imp

PLATE XII.

Bowmanites Cambrensis, Binney.

Fig. 1 (No. 30).   Natural mould of the Stem, Branch, and lower portion of the Cone, containing Macrospores, in a nodule of clay-ironstone, from the Varteg Iron-works, South Wales.   Natural size.

Fig. 1*a*.   Natural mould of a portion of the Stem and two whorls of Leaves, in another piece of the nodule.   Magnified 2 diameters.

Fig. 1*b*.   A cast (in gutta percha) of a portion of the Stem, with Joints, Knots, and Leaves.   Magnified 2 diameters.

Fig. 1*c*.   A cast (in gutta percha) of the Branch to which the Cone was attached. Magnified 2 diameters.

Fig. 2.   The Stem, Branch, Leaves, and Cone, as restored by the late Mr. Bowman. Natural size.

Fig. 3.   Three of the lower Scales or Bracts, two of them supporting each five Macrospores, from Mr. Bowman's original sketch.   Magnified 6 diameters.

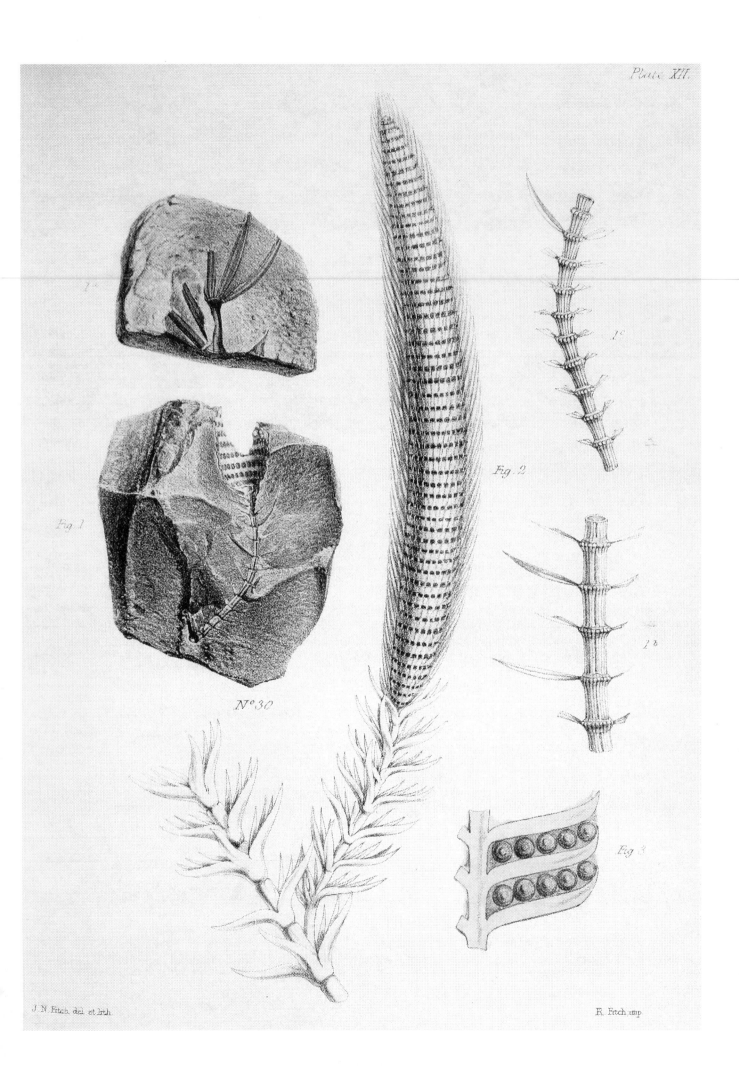

Plate XII.

1ᵃ

Fig. 2

1ᶜ

Fig. 1

1ᵇ

Nº 30

Fig. 3

J. N. Fitch, del. et lith.

R. Fitch, imp.

THE

PALÆONTOGRAPHICAL SOCIETY.

INSTITUTED MDCCCXLVII.

VOLUME FOR 1871.

LONDON:
MDCCCLXXII.

OBSERVATIONS

ON THE

STRUCTURE OF FOSSIL PLANTS

FOUND IN THE

CARBONIFEROUS STRATA.

BY

E. W. BINNEY, F.R.S., F.G.S.

PART III.

LEPIDODENDRON.

PAGES 63—96; PLATES XIII—XVIII.

LONDON:
PRINTED FOR THE PALÆONTOGRAPHICAL SOCIETY.
1872.

PRINTED BY

J. E. ADLARD, BARTHOLOMEW CLOSE.

CONTENTS.

		PAGE
I. INTRODUCTORY REMARKS	. . . .	63
II. BIBLIOGRAPHICAL HISTORY OF LEPIDODENDRON	. . . . .	64
§ 1. Adolphe Brongniart	. . . .	64
2. Carruthers	. . . .	65
3. Schimper	. . . .	70
4. Williamson	. . . .	71
III. GENERAL OBSERVATIONS ON LEPIDODENDRON, STIGMARIA, AND HALONIA	. . .	74
IV. DESCRIPTION OF THE SPECIMENS	. . . . . .	77
§ 1. Specimen No. 31. *Lepidodendron Harcourtii*, Pl. XIII, figs. 1—6	. . .	77
2. Specimen No. 32. — — Pl. XIV, figs. 1—3	. .	80
3. Specimen No. 33. *Sigillaria vascularis*, Pl. XIV, figs. 4—6	. . .	81
V. BIBLIOGRAPHICAL HISTORY OF HALONIA	. . . . . .	82
§ 1. Lindley and Hutton	. . . .	82
a. *Halonia ? tortuosa*	. . . . .	82
b. — *gracilis*	. . . . .	83
c. — *regularis*	. . . . .	83
2. Richard Brown	. . . .	83
3. Unger	. . . .	84
4. Dawes	. . . .	84
5. Hooker	. . . .	86
6. Brongniart	. . . .	86
7. Goldenberg	. . . .	87
8. Binney	. . . .	88
9. Schimper	. . . .	88
VI. DESCRIPTION OF SPECIMENS OF HALONIA	. . . . .	89
§ 1. Specimen No. 34. *Halonia regularis*, Pl. XV, figs. 1—4	. . .	89
2. Specimen No. 35. — — Pl. XVI, figs. 1—5	. . .	90
3. Specimen No. 36. — — Pl. XVII, figs. 1, 2, 3	. .	92
4. Specimen No. 37. — — Pl. XVII, figs. 4, 5, 6	. .	93
5. Specimen No, 38. — — Pl. XVIII, fig. 1	. . .	94
VII. CONCLUDING REMARKS	. . . .	95

PART III.

I. INTRODUCTORY REMARKS.

IN the Part last published of this Monograph a deviation was made from the plan first proposed, in considering some of the Organs of Fructification of Fossil Plants, and leaving the treatment of the structure of the Stems for a time. I will now, therefore, return to the original intention of describing the internal structure of several specimens of Fossil Woods from my own cabinet and that of Mr. J. S. DAWES. These are in a good state of preservation, and will afford some additional information on the genus LEPIDODENDRON.

After the investigations of Witham, Lindley and Hutton, Brongniart, Hooker, and others, nearly every portion of the structure of the stem had been made out, with the exception of the medulla. It was more inferred than proved that the structure of this part consisted of cellular tissue in the Rev. C. G. W. HARCOURT's specimen found at Hesley Heath. Mr. Dawes' specimen, however, figured in PLATE VII, and described in PART II of this Monograph, clearly proves that the above-named authors were in the main right in stating that the medulla of *Lepidodendron* was composed of a cellular tissue which was not ordinary parenchyma, like such as is found immediately outside the vascular cylinder; for it consisted of large rectilinear cells, arranged in vertical series, something like bricks upon their ends, as first noticed by Mr. Dawes to occur in *Halonia*. This beautiful *Lepidodendron*, probably the most perfect in its state of preservation of any with which we are acquainted, will be described in detail before I proceed to consider the genus *Halonia*, and to show the relation of that fossil to *Lepidodendron*. Some remarkable instances of the division of the vascular axes of *Lepidodendron, Halonia*, and *Sigillaria vascularis*, just before the stems of those plants dichotomize, will be given. These fully confirm Brongniart's views, as stated in his admirable memoir on *Les Lycopodiacées*, in the second volume of his 'Histoire des Végétaux fossiles.'

I shall then restate the chief part of what has been published on the structure of *Halonia*. Next a description of Mr. Dawes' and my own specimens will be given in detail. In conclusion, it is intended to show the connection of *Halonia* with *Lepidodendron*, so far as their structure is concerned.

10

The *Halonia regularis* of Lindley and Hutton is the only species that will be now described, as the other species, although similar in structure, vary very much in their external characters from that species.

II. Bibliographical History of Lepidodendron.

§ 1. At page 34 of Part II of this Monograph reference is made to some of the authors who had written on *Lepidodendron*; but since the time of the publication of M. Adolphe Brongniart's memoir on *Lycopodiacées*[1] probably little has been done to increase our knowledge of the structure of the stem of *Lepidodendron*. The restored figure in his plate xxi is nearly similar in structure to that of the specimen now about to be described, with the exception of the medulla or pith and the tissue enclosing the vascular bundles, which in the specimen from Hesley Heath were not very well preserved, and therefore not altogether correctly shown. To those who desire to obtain a knowledge of the structure of *Lepidodendron* and its relation to recent *Lycopodiaceæ*, a reference to the work itself is necessary. For the present probably it will be sufficient for my purpose to quote the author's views, as expressed at page 45 of his work.

" Si, après avoir ainsi cherché à déterminer les rapports de position des divers tissus qui entrent dans la composition de ce rameau de Lépidodendron, nous examinons la structure même de chacun de ces tissus, et surtout de quelques-uns d'entre eux, nous verrions que la nature des fibres vasculaires qui composent soit le cylindre central, soit les faisceaux qui se portent dans les feuilles, confirme tout à-fait l'analogie des ces plantes avec les Lycopodiacées. Ainsi, le cylindre vasculaire qui entoure le tissu cellulaire central, et qui occupe presque le centre de la tige est uniformément composé du même tissu ; il n'y a pas mélange de fibres et de vaisseaux de diverses sortes, comme dans les faisceaux fibro-vasculaires de Phanérogames ; il est, au contraire, entièrment formé d'élémens identiques par leur structure, et qui ne diffèrent que par leur volume et la déformation que la compression leur a donnée, disposition tout-à-fait semblable à celle qu'on observe dans les faisceaux vasculaires aplatis des Lycopodiacées et dans ceux des Fougères. Enfin, ces fibres ont cette structure toute particulière, si habituelle dans les végétaux de cette classe naturelle, qu'on a désignée sous le nom de vaisseaux rayés, vaisseaux fendus, vaisseaux scalariformes (treppengange), et qui paraissent former un des caractères essentiels de ces végétaux. Le volume beaucoup moindre de ces vaisseaux dans la partie extérieure du cylindre vasculaire et dans les faisceaux qui s'en séparent pour se porter dans les feuilles, est encore un caractère qu'on observe généralement dans

---

[1] ' Histoire des Végétaux fossiles,' Tome Deuxième (published in 1837).

les Lycopodiacées, et particulièrement dans celles dont les feuilles très nombreuses, sont en rapport avec un nombre proportionnel de faisceaux vasculaires entourant l'axe central. (Voyez la coupe de la tige du *Lycopodium verticillatum*, pl. x, fig. 1.)

" Enfin, la délicatesse du tissu cellulaire qui environne extérieurement ce cylindre vasculaire, et qui le sépare du tissu cellulaire plus dense qui forme la zone résistante extérieure de la tige, la destruction facile de ce tissu, la position excentrique de l'axe vasculaire dans l'espèce de cavité cylindrique qui résulte de la destruction plus ou moins complète de ce tissu cellulaire, sont des caractères ou des dispositions accidentelles qu'on recontre très fréquemment sur les tiges sèches des Lycopodiacées conservées dans nos herbiers.

" Ainsi, par la structure intérieure de leurs tiges, comme par leur forme extérieure, leur mode de ramification et la disposition de leurs feuilles, les Lépidodendrons s'accordent presque complètement avec les Lycopodiacées, et ne seraient autre chose que des Lycopodiacées arborescentes."

§ 2. *Carruthers.*—This author, in his paper published in vol. xiii, pp. 2 to 9, ' Journ. Botany,' says " *Lepidodendron* was a branching tree of considerable size. It is separated from the other genera of coal plants by the form and arrangements of the leaf-scars upon its stem. More than forty species have been recorded ; but, as the scars present different appearances on different portions of the same plant, no doubt more species have been established than the materials fairly warrant. But that they were numerous in species, and very numerous in individuals, any one who has even cursorily examined a coal-pit, or the fossils in any public museum, must be convinced. They certainly contributed largely to the formation of coal.

" The researches of Witham, Lindley and Hutton, Brongniart, and Binney have made us acquainted with the stem."

Mr. Carruthers, in a note referring to my papers on *Sigillaria vascularis*, published in the ' Quarterly Journal of the Geological Society,' and in the ' Phil. Transactions,' speaking of myself, says, " He refers them to the genus *Sigillaria*, because of their agreement in internal structure with Brongniart's *Sigillaria elegans ;* but he cannot separate them by their external markings from *Lepidodendron selaginoides*, Lindl. and Hutt. ; and, as the only characters by which the two genera are distinguished are derived from the markings of the stem, we must consider *Sigillaria vascularis* as a true *Lepidodendron*. I am the more satisfied as to this because I believe no essential difference exists, as has been hitherto maintained, between the stems of *Sigillaria* and *Lepidodendron*, or any of the other lepidodendroid plants of the coal period. I cannot enter into this question here, but I shall take an early opportunity of publishing my views and the reasons for maintaining them."

Mr. Carruthers then goes on to state, " These published observations, together with the examination of some beautiful specimens in the collection of Robert Brown, now in the Botanical Department of the British Museum, and of Mr. Alexander Bryson, enable me to give a somewhat complete description of its singular structure.

"The axis of the stem cannot be considered as a true medulla or pith, inasmuch as it is composed, not of simple cells, but of elongated utricles, of various sizes, irregularly arranged, and having their walls marked with scalariform bars (pl. lvi, fig. 2). These utricles indeed differ from the vascular tissue of the woody cylinder which surrounds them, only by their length. The tissue of the woody cylinder consists of long scalariform vessels, which increase in size from the inner margin to the outer; this increase being sufficient to meet the requirements of the enlarged circumference, with the help of only a few additional series of vessels. As there is no true medullary cellular tissue in the axis, so there are no medullary rays passing through this cylinder. In radial sections an appearance is seen singularly resembling, to the naked eye, the silver grain produced in dicotyledonous woods by the medullary rays, but this arises from a very different cause. The diameter of the vessels is so great that, on a polished surface, only the scalariform wall of the vessel that lies on or near the surface is exhibited; and when the upper wall of a vessel is cut away, the lower wall is often so deeply buried in the opaque substance that the peculiar structure is obscured. In the case of sections prepared for microscopic examination, both surfaces of some vessels are often removed, and the scalariform markings on the lateral walls, or on any horizontal walls which by chance occupy a medial position between the polished surfaces, only are seen. This absence of the scalariform bars gives at first sight the appearance produced by medullary rays.

"The continuous cylinder of scalariform vascular tissue appears to be penetrated by the vascular bundles which ultimately supply the leaves. These bundles apparently originate either in the scalariform tissue of the axis or on the inner surface of the woody cylinder. They have been mistaken for, or misnamed, medullary rays.

"The woody cylinder is surrounded by a great thickness of cellular tissue, which extends to the exterior of the stem, and is composed of three distinct and separate zones. The inner zone has never, so far as I know, been perfectly preserved in any specimen, yet traces of it sometimes may be seen, and it is rightly restored in Brongniart's drawing of *Lepidodendron Harcourtii*, in the 'Archives du Muséum,' vol. i, pl. xxxi. Its absence in fossils is owing to its extremely delicate structure. The cells of the middle zone have thicker walls, and they have consequently frequently resisted decomposition before fossilization made them permanent. In the outer zones the cells are very much lengthened, and have a smaller diameter. They nearly resemble true vascular tissue; but the progress of lengthening may easily be traced from the interior outwards, and no distinction can be drawn between the true cells and the long and slender ones of the outer circumference. The cell-walls of all the three zones are without markings of any kind.

"These three cellular zones are traversed by the vascular bundles which rise from the outside of the interior woody cylinder, if they do not actually pass through it, and pass to the leaves and branches. These bundles separate from the woody cylinder a long way below the point where they pass off into the leaf. At first their direction is almost parallel with the cylinders, slightly inclining outwards ; they then incline more outwards ; and as they approach the circumference of the stem, they resume their nearly ascending direction for some distance, until they finally pass out to the leaves which they support. Each bundle consists of scalariform vessels, very much finer than those of the woody cylinder, surrounded by elongated cells, like those of the outer zones, and probably still further enclosed by a delicate parenchyma, which disappeared before it could be fossilized. The only evidence I have of the existence of this cellular tissue is that the bundles never fill the cavities in the parenchyma of the stem through which they pass. The bundles terminate in the points seen on the areolas of the stem, which are the scars of the leaves.

"The woody cylinder is of different thicknesses in different stems, and appears to have increased with the growth of the tree. There is, however, no indication of interruption in the growth, or of seasonal layers. Yet it cannot be conceived that the whole vascular cylinder arose and was developed at the same time. It is very probable that the zone of slender and consequently of rarely preserved cellular tissue, which surrounds the woody cylinder, was analogous in its functions to the cambium layer of phanerogamous stems, like the similar layers in recent *Lycopodiaceæ,* described by Spring in his ' Monographie de la Famille des Lycopodiacées' (p. 294).

" If we separate the different structures we have described in the axis into two series, the one series axial, and the other epidermal, we shall have the axis composed of scalariform utricles, the woody cylinder and the vascular bundles passing to the leaves belonging to the first series, and the two external zones of the vascular tissue to the second. The inner zone of cellular tissue, like the cambium layer, was most probably common to both series, the cells of the outer circumference being developed into the parenchyma of the epidermal series, while the vessels of the woody axis were produced from the cells of the inner series."

Here come in the paragraphs on the Strobilus of *Lepidodendron* and its stigmarioid roots, quoted by me at length in pages 37 and 38 of Part II of this Monograph. Then the author proceeds, " In speculating upon the conditions under which the forests of *Lepidodendron* flourished, it is most important to observe whatever is peculiar in those organs by which the plants were connected with the physical conditions around them. Geologists have too much overlooked such considerations in their deductions as to the physical phenomena of a period from the plants and animals that then existed. They have often taken for granted that the known conditions of the living species of a genus are true also of the fossil members of the same genus. In the want of other evidence such an assumption may be cautiously employed ; but, unless its true value be accurately estimated, the greatest errors may arise, as they have in the past. For example, the systematic position of the *Elephas primigenius* having been clearly established, the inference was thought legitimate that, as the modern representatives of the genus were confined to tropical or subtropical countries, the boreal regions must have enjoyed a similar climate when they were inhabited by these ancient elephants. It was, however, discovered that their skin was clothed with wool and long hair, and that consequently they were adapted to endure a cold climate. In plants the structure of the fruit would, in most cases, teach nothing as to the temperature and humidity of the atmosphere in which, or the kind of soil upon which, the organism grew, though it would be of the first importance in determining the systematic position. On the other hand, the root, the leaves, and the tissues of the plant would be of only secondary importance in regard to systematic position, but of the highest value in determining physical condition. In regard to *Lepidodendron*, its singular roots would seem to imply that it derived a large amount of moisture through them from a moist soil, and so far differed from most living Cryptogamia which obtain it mostly from the atmosphere. The roots of this genus presented in their crowded and long rootlets an immense surface for the absorption of moisture; and, in their great abundance of lax cellular tissue, possessed the means of containing this moisture and transmitting it to the foliage.

"The leaves of *Lepidodendron* were simple, lanceolate, acute, and sessile. They had a single medial nerve. The younger branches were densely crowded with leaves; and the scars left on the trunk, after they perished, give the numerous beautiful markings by which the species have been distinguished. The leaves, when found separated from the branches, are called *Lepidophylla*. . . . .

. . . "The structure of the arboreal stem of *Lepidodendron* is much more complex than that of any known cryptogam. The central axis of the irregularly arranged vascular tissue in *Lycopodium* is suited to the low stature of the plants of that genus; but in the giant *Lepidodendron* there is a complexity which approaches the structure of some dicotyledonous stems. The general arrangement of the tissues, resembling what exists in some *Cycadeæ* and *Cactaceæ*, has caused this fossil plant to be referred sometimes to the one and sometimes to the other of these Orders; but the resemblance is only one of analogy and not of affinity. The presence of scalariform tissue, of which the woody portion is entirely composed, and the absence of medullary rays, would, even if the fruit were unknown, be sufficient to establish the cryptogamic nature of the plant. A comparison with the Cycadean stem may help us, by the resemblances and differences which will appear, better to understand the stem of *Lepidodendron*. The Cycads have all a large medulla, composed of large-sized parenchyma, in some genera traversed by numerous vascular bundles, as in *Encephalartus*, and in others entirely cellular, as in *Cycas* and *Zamia*. This is surrounded by a single woody cylinder, or several, everywhere penetrated with medullary rays. Beyond this there is a considerable thickness of parenchyma, composed in their inner portion of cells whose length exceeds only slightly their breadth; these gradually lengthen, until they assume an appearance very like the external portion of *Lepidodendron*. This cortical parenchyma is traversed by the vascular bundles which supply the leaves. The two stems are evidently built upon the same plan; and were we to substitute scalariform tissue for the gymnospermatous woody tissue, and scalariform utricles for true medullary parenchyma, and finally exclude the medullary rays, the description of the Cycadean stem would apply to that of *Lepidodendron*. And it deserves special notice, that the surface of the Cycadean trunk is composed of the bases of the old leaves, together with the scales which in some species are interspersed among them or alternate with them. The leaves do not disarticulate at the circumference of the stem, but at some distance from it, leaving a small portion of the base persistent. The scars of the outer surface of the stem give a different impression from those presented when the persistent bases of the leaf-stalks are removed. Whoever is even a little familiar with coal fossils is aware that there are two sets of scars on the stems of *Lepidodendron*—one superficial and the other internal. The fossils that present the first set are generally said to be 'corticated' stems, and those exhibiting the others 'decorticated.' The 'bark' is generally converted into a compact structureless coal, the outer surface of which has the one set of scars, and the inner surface the other. I believe this coal is produced by the external of the two epidermal series, and that the outer scars were truly superficial, while the inner were produced by the vessels which passed to the bases of the leaves. The two sets of scars in Cycadean stems are analogous structures ; but in *Lepidodendron* the layer which bears the scars on its two surfaces is a compact cylinder ; while in *Cycadeæ* there is no connecting tissue uniting the bases of the leaves ; they are closely packed together, but quite free from each other. It is evident that in many respects the fossil stem had a striking analogy in the arrangement of its parts to that of the recent Cycads, while it was a true Cryptogam ; and if we now examine the slender stem of *Lycopodium*, we shall find, I believe, that *Lepidodendron*, though more highly developed, does not differ essentially from it.

"Spring, in his 'Monographie des Lycopodiacées' (p. 293), describes the stem of this Order as composed of five parts :—1st, the woody axis ; 2nd, a layer of delicate cells ; 3rd, the liber ; 4th, the herbaceous envelope ; and, 5th, the epidermis.

"The axis is composed of bundles of scalariform vessels, scattered through a very delicate cellular tissue, in a regular figure, which varies in the different species. This axis is surrounded by a layer of lax, delicate, cellular tissue, which Spring considers to be the channel through which the sap circulates, and the seat of growth in the stem, the inner portion being developed into wood vessels, and the outer into 'liber.' The 'liber' is composed of elongated cells, with thickened walls. Spring gives to it this name because of its analogy to the liber in dicotyledons. This layer is often so thin that it is difficult to detect. It is surrounded by a thick greenish layer, composed of large elongated cells with thin walls ; and this is covered with an epidermis, consisting of small cells with thick walls. The vascular bundles pass through the various layers of cellular tissue from the axis to the circumference.

"The great difference between the stem of *Lepidodendron* and *Lycopodium* is the existence of a pseudo-medulla and the arrangement of the vascular tissue as a solid cylinder in the fossil genus, compared with the central position and loose structure of the vascular tissue in the recent plant. In both the recent and fossil stems the vascular tissues are surrounded by a zone of thin-walled cells, which has disappeared in all the dried specimens of *Lycopodium* I have examined, having the axis free, and which, as we have seen, is very rarely preserved in *Lepidodendron*."

§ 3. *Schimper.* Professor Schimper[1] describes his Family of *Lepidodrendeæ* as follows :—*Plantæ arborescentes, foliis homomorphis, lanceolatis et linearibus, plano-carinatis, integerrimis, spiraliter dispositis, deciduis, cicatricesque regulares relinquentes ; trunci fasciculis vascularibus in cylindrum continuum conjunctis solum parenchyma medullare continente vel parenchyma vasis intermixtum ; fructificatione strobiliformi, sporangia elongata bractearum basi horizontali adfixa lateraliter dehiscentia gerente.*

" Les feuilles articulées caduques laissant des cicatrices régulières persistantes pendant tout la durée de la plante, le cylindre vasculaire continu ne refermant que du parenchyme médullaire ou formé entièrment par un tissu fibro-vasculaire, les strobiles caducs portant des sporanges allongés placés horizontalement sur la base des bractées, sont des caractères distinctifs qui paraissent autoriser l'établissement d'une famille spéciale pour ce groupe de végétaux fossiles formé de plusieurs genres distincts.

" Toutes le Lepidodendrées paraissent avoir été arborescentes. On en a trouvé des troncs qui avaient une longueur de plus de 100 pieds et un diamètre de 10 à 12 pieds. Le tronc était simple jusqu'à hauteur considérable ; la ramification se faisait par dichotomie régulière ; les feuilles persistaient assez longtemps sur les rameaux, comme dans nos Conifères ; et leurs cicatrices, quoique changées à la suite du développement de la plante, conservaient cependant une forme très-régulière, même sur les troncs les plus âgés."

The same author, after describing the outside appearance of *Lepidodendron,* STERNB., *Sagenaria,* BRONG., at p. 16, says, " M. Brongniart compare la structure interne du tronc des Lépidodendrons à celle de la tige des *Psilotum* et des *Tmesipteris.* Le cylindre vasculaire de l'espèce examinée par ce savant, du *Lep. Harcourtii,* est en effet, comme dans ces deux genres, simple et sans lacunes médullaires il est formé intérieurement de larges vaisseaux scalariformes, et extérieurement d'étroits vaisseaux rayés (spiralés ?), d'où partent des faisceaux composés de fibres ou de vaisseaux tout-à-fait semblables à ces derniers, pour se rendre dans les feuilles en décrivant un arc qui se rapproche de l'horizontale vers son extrémité. Le cylindre lui-même est renfermé dans une gaîne épaisse d'un tissu parenchymateux très-solide qui est suivi du parenchyme cortical, également épais, mais d'une texture plus lâche. Dans ce dernier tissu, on remarque deux zones d'un aspect un peu différent, et dans lesquelles Corda a cru distinguer le liber et le parenchyme cortical des Phanérogames. Un tissu superficiel, enfin, composé de cellules très-étroites et allongées constitue l'epiderme. Les Lycopodes vivants montrent du reste exactement les mêmes

---

[1] 'Traité de Paléontologie végétale,' Tome second, Première Partie, p. 13, 1870.

differences dans les diverses couches du tissu parenchymateux qui se suivent depuis la surface du cylindre vasculaire jusqu'à l'épiderme inclusivement.

"Il est à noter cependant que ce qui a été dit sur la structure interne ne se rapporte qu'à une seule espèce, le *Lep. Harcourtii*, prise pour type, et il n'est pas du tout certain que tous les fossiles qui, à la suite de leur organisation extérieure, se trouvent réunis dans ce genre, aient aussi la même organisation intérieure ; cela est même peu probable. M. Binney, dans son mémoire cité plus haut (*Sigillaria and Lepidodendron*, printed in the 'Proceedings of the Geological Soc. of London,' Jan., 1862), parle de deux plantes dont le cicatrices foliaires coincident parfaitement avec celles des *Lepidodendron*, et dont l'une, le *Sigillaria vascularis*, offre la structure des Sigillaires, tandis que l'autre, le *Lepidodendron vasculare*, a son axe central entièrement composé de gros vaisseaux scalariformes et de fins vaisseaux rayés, au lieu d'offrir un cylindre vasculaire occupé intérieurement par un parenchyme médullaire, comme dans le *L. Harcourtii*. Ce type se rapprocherait donc davantage des Lycopodes, et le dernier des *Psilotum* et *Tmesipteris*."

§ 4. *Williamson.*—Professor W. C. Williamson, F.R.S., in a paper read before the Royal Society, June 15, 1871,[1] "On the Organisation of the Fossil Plants of the Coal-Measures, Part 2, *Lepidodendron* and *Sigillaria*," says, " The *Lepidodendron selaginoides* described by Mr. Binney, and still more recently by Mr. Carruthers, is taken as the standard of comparison for numerous other forms. It consists of a central medullary axis composed of a combination of transversely barred vessels with similarly barred cells ; the vessels are arranged without any special linear order. The tissue is closely surrounded by a second and narrower ring, also of barred vessels, but of smaller size, and arranged in vertical laminæ which radiate from within outwards. These laminæ are separated by short vertical piles of cells, believed to be medullary rays. In the transverse section the intersected mouths of the vessels form radiating lines, and the whole structure is regarded as an early type of an exogenous cylinder ; it is from this cylinder alone that the vascular bundles going to the leaves are given off. This woody zone is surrounded by a very thick cortical layer, which is parenchymatous at its inner part, the cells being without definite order ; but externally they become prosenchymatous, and are arranged in radiating lines, which latter tendency is observed to manifest itself whenever the bark-cells assume the prosenchymatous type. Outside the bark is an epidermal layer separated from the rest of the bark by a thin bast-layer of prosenchyma, the cells of which are developed into a tubular and almost vascular form ; but the vessels are never barred, being essentially of the fibrous type. Externally to this bast-layer is a more superficial epiderm of parenchyma supporting the bases of the leaves, which consist of similar parenchymatous tissue. Tangential sections of these outer cortical tissues show that the so-called 'decorticated' specimens of *Lepidodendron*, and of other allied plants, are merely examples

---

[1] 'Proceed. Roy. Soc.,' vol. xix, p. 500, &c. ; 'Nature,' No. 87, vol. iv, p. 173, June 29th, 1871.

that have lost their epidermal layer, or had it converted into coal; this layer, strengthened by the bast-tissue of its inner surface, having remained as a hollow cylinder, when all the more internal structures had been destroyed or removed.

" From this type the author proceeds upwards, through a series of examples in which the *vessels* of the medulla become separated from its central *cellular* portions and retreat towards its periphery, forming an outer cylinder of medullary vessels, arranged without order, and enclose a defined cellular axis; at the same time the encircling ligneous zone of radiating vessels becomes yet more developed, both in the number of its vessels and in the diameter of the cylinder relatively to that of the entire stem. As these changes are produced, the medullary rays separating the laminæ of the woody wedges become more definite, some of them assuming a more composite structure, and the entire organization gradually assuming a more exogenous type. At the same time the cortical portions retain all the essential features of the Lepidodendroid plants.

" We are thus brought by the evidence of internal organisation to the conclusion that the plants which Brongniart has divided into two distinct groups, one of which he has placed amongst the Vascular Cryptogams, and the other amongst the Gymnospermous Exogens, constitute one great natural family. . . . . . .

" *Stigmaria* is shown to have been much misunderstood, so far as the details of its structure are concerned, especially of late years. In his memoir on *Sigillaria elegans*, published in 1839, M. Brongniart gave a description of it, which, though limited to a small portion of its structure, was as far as it went a remarkably correct one. The plant, now well known to be a root of *Sigillaria*, possessed a cellular pith without any trace of a distinct outer zone of medullary vessels, such as is universal amongst the *Lepidodendra*. The pith is immediately surrounded by a thick and well-developed ligneous cylinder, which contains two distinct sets of primary and secondary medullary rays. The primary ones are of large size, and are arranged in regular quincuncial order; they are composed of thick masses of mural cellular tissue. A tangential section of each ray exhibits a lenticular outline, the long axis of which corresponds with that of the stem; these rays pass directly outwards from pith to bark, and separate the larger woody wedges, which constitute so distinct a feature in all transverse sections of this zone, and each of which consists of aggregated laminæ of barred vessels, disposed in a very regular radiating series. The smaller rays consist of vertical piles of cells arranged in single rows, and often consisting of but one, two, or three cells in each vertical series; these latter are very numerous, and intervene between all the numerous radiating laminæ of vessels that constitute the larger wedges of woody tissue. The vessels going to the rootlets are not given off from the pith, as Goeppert supposed, but from the sides of the woody wedges bounding the *upper* part of the several large lenticular medullary rays; those of the *lower* portion of the ray taking no part in the constitution of the vascular bundles. The vessels of the region in question descend vertically, and parallel to each other, until they come in contact with the medullary ray, when they are suddenly deflected, in large numbers, in an outward direction, and nearly at right angles to their previous course, to reach the rootlets. But only a small number reach their destination, the great majority of the deflected vessels terminating in the woody zone. A very thick bark surrounds the woody zone. Immediately in contact with the latter it consists of a thin layer of delicate, vertically elongated, cellular tissue, in which the mural tissues of the outer extremities of the medullary rays become merged. Externally to this structure is a thick parenchyma, which quickly assumes a more or less prosenchymatous form, and becomes arranged in thin radiating laminæ as it extends outwards. The epidermal layer consists of cellular parenchyma, with vertically elongated cells at its inner surface, which feebly represents the bast-layer of the other forms of Lepidodendroid plants. The rootlets consist of an outer layer of parenchyma, derived from the epidermal parenchyma. Within this is a cylindrical space, the tissue of which has always disappeared. In the centre is a bundle of vessels surrounded by a cylinder of very delicate cellular tissue, prolonged either from one of the medullary rays, or from the delicate innermost layer of the bark, because it always accompanies the vessels in their progress through the middle and outer barks.

"The facts of which the preceding is a summary lead to the conclusion that all the forms of plants described are but modifications of the Lepidodendroid type. The leaf-scars of the specimens so common in the coal-shales represent tangential sections of the petioles of leaves when such sections are made close to the epidermal layer. The thin film of coal of which these leaf-scars consist, in specimens found both in sandstone and in shale, does not represent the entire bark as generally thought, and as is implied in the term 'decorticated' usually applied to them, but it is derived from the epidermal layer. In such specimens all the more central axial structures, viz., the medulla, the wood, and the thick layer of true bark, have disappeared through decay, having been either destroyed, or in some instances detached and floated out ; the bast-layer of the epiderm has arrested the destruction of the entire cylinder, and formed the mould into which inorganic materials have been introduced. On the other hand, the woody cylinder is the part most frequently preserved in *Stigmaria;* doubtless because, being subterranean, it was protected against the atmospheric action which destroyed so much of the stem.

"It is evident that all these Lepidodendroid and Sigillarian plants must be included in one common family, and that the separation of the latter from the former as a group of Gymnosperms, as suggested by M. Brongniart, must be abandoned. The remarkable development of exogenous woody structures in most members of the entire family indicates the necessity of ceasing to apply either to them or to their living representatives the term Acrogenous. Hence the author proposes a division of the vascular cryptogams into an Exogenous group, containing *Lycopodiaceæ, Equiseticeæ*, and the fossil *Calamitaceæ*, and an Endogenous group, containing the Ferns,—the former uniting the Cryptogams with the Exogens through the *Cycadeæ* and the other Gymnosperms, and the latter linking them with the Endogens through the *Palmaceæ*."

III. GENERAL OBSERVATIONS ON LEPIDODENDRON, STIGMARIA, AND HALONIA.

The genus *Lepidodendron*, as shown by the Hesley Heath and Dudley specimens, although they are very similar in their external characters with small specimens of *Sigillaria vascularis*, differs much from them in its internal structure. The first-named plant shows no trace of a radiating cylinder of barred tubes arranged in wedge-shaped masses, or anything like medullary rays, whereas the last-named affords evidence of both these important characters, and in those parts it is identical in structure with *Stigmaria*. The specimens figured and described by me in the 'Quarterly Journal of the Geological Society,' and in the 'Philosophical Transactions,' amply prove that *Sigillaria* was an exogenous tree, and had an internal vascular cylinder arranged in wedge-shaped masses, and penetrated by medullary rays, as well as an external zone arranged in radiating series of strong vascular-like tubes, or utricles, without any appearance of bars or discs, and penetrated by large vascular bundles. Both the large and small specimens were the same in internal structure; and the one could be shown to pass gradually into the other, the rhomboidal scars of the former giving place to the irregular ribs and furrows of the latter.

Professor Williamson has had in his possession for many years a specimen of *Stigmaria*, figured and described by me in 1849,[1] which, although the medulla is absent, distinctly shows in the inside of the cylinder, as left, the medullary rays traversing that portion of the stem. Several specimens in my cabinet also distinctly prove the existence of these vessels in *Stigmaria*.

Sir William Logan, myself, and others have alluded to the occurrence of *Stigmaria* in the floors of coal-seams. The specimens of this root there found are almost always in a very indistinct and distorted condition, and the only characters generally to be seen are the impressions or casts of some circular areolæ and the traces of rootlets in the fire-clay. Yet from these imperfect characters the whole of *Stigmariæ* in this country have been generally included in the species *ficoides*. It has long been my opinion that the external characters of this root are common to several other genera of aquatic plants besides *Sigillaria*, and that such roots would have a great similarity in appearance from their former soft and muddy *habitats*. Now, all the specimens of *Stigmaria* which I have seen, and whose connection I have traced with upright stems of *Sigillaria*, have always been of a comparatively large size, seldom if ever less than from two to three inches in diameter. The *Sigillaria* at its base divides into four main roots, each of which bifurcates into two secondary roots, which in their turn again dichotomize into two tertiary roots. Each of these last present on their surface an irregularly ribbed and furrowed, or, more correctly speaking, a gnarled appearance for a considerable distance outwards; and this gradually

'Quart. Journ. Geol. Soc.,' vol. vi, p. 17.

disappears, and is succeeded at first by indistinct holes of a circular form, or projecting circular knobs, which soon pass into distinct areolæ with an elevated point in their centre, and radicles radiating from the root in all directions. When this state of the root is attained, the gradual decrease in bulk, which has gone on from the base of the *Sigillaria* to this point, ceases, and the *Stigmaria* root runs to the length of twenty feet and upwards with about the same size. Its absolute termination is not often seen ; but, from specimens found, it is generally believed to have an obtuse club-shaped end. Certainly I have never seen it dichotomize into smaller roots, although other observers may have been more fortunate.

Professor Williamson, in his ' Observations' previously quoted, when speaking of the structure of the medulla of *Stigmaria*, says that it possessed " a cellular pith without any trace of a distinct outer zone of medullary vessels, such as is universal amongst the *Lepidodendra*." I have no doubt that he has found in numerous small stems, obtained from the "Upper Foot Coal" near Oldham (a locality which I first observed and pointed out to others, but from which time I have never been able to obtain any more specimens for myself), evidence of the cellular pith he alludes to. This cellular tissue is not the ordinary parenchyma usually forming piths, but the " orthosenchymatous" tissue common as a medulla to *Lepidodendron, Halonia*, and *Calamodendron* (?). This structure (in the locality above named) is found to prevail in many stems possessing radiating woody cylinders of barred tubes, penetrated by medullary rays, and varying in diameter from one twentieth of an inch to two inches. All these specimens, from the similarity of their structure, might be taken to be *Stigmaria ;* but we seldom find their external characters so well preserved as to recognise them as identical with the large roots found connected with *Sigillaria* stems. From specimens in my possession it is almost certain that there are fossil roots, having all the outward characters of *Stigmaria*, some of which have their medulla composed of the cellular tissue alluded to by Professor Williamson ; whilst others have a medulla composed of a combination of transversely barred vessels and cellular tissues, or what I have ventured to term " orthosenchymatous" tissue, as shown in my specimens of *Sigillaria vascularis* and the *Stigmaria* figured and described by Professor Geoppert. It is also probable that these roots may have belonged to distinct, but allied, trees. These two different kinds of medulla appear to have existed in Lepidodendroid trees ; the former in *Lepidodendron Harcourtii*, and the latter in my *Lepidodendron vasculare*.

It is scarcely necessary here to allude to the great number of genera and species of fossil plants which have been erected on imperfect and ill-preserved specimens, only exhibiting some of their external characters ; but even amongst these, few persons could determine a compressed *Stigmaria ficoides* from a *Halonia regularis*, except where the latter was observed to dichotomize. When we examine the structure of the vascular bundles traversing the stem from the central woody axis to the leaves or rootlets, we have difficulty in distinguishing those which belong to *Sigillaria* and *Lepidodendron* from those of *Stigmaria* and *Halonia*, only the latter are considerably larger. In fact the *Stigmaria*, from the structure of its rootlet, might be supposed to be a stem (as originally entertained) rather than a root, as it has undoubtedly been proved to be by actual connection of the one with the other.

When we come to examine more carefully the specimens of plants found in coal-floors, it is probable that we shall find the remains of the roots of many different plants besides those of *Sigillaria*. The greatest difficulty we have to encounter in our investigation is the wretched condition in which the fossils are found, being usually very much compressed and distorted; and seldom in my own experience, except in the case of "Gannister" floors, has it been possible to obtain any trace of the internal structure of the plants.

It must always be borne in mind that the specimens of fossil woods showing structure in a perfect condition are generally of a very small size; those of *Calamodendron commune*, and the small *Stigmaria*-like plant from the Upper Foot Coal, often being only from one twentieth to one thirtieth of an inch in diameter. In such small spaces every cell of the medulla, as well as those in the radiating woody cylinder, is preserved in as perfect a condition as it existed in the living plant. But *Sigillaria vascularis*, *Lepidodendron*, and *Halonia* are seldom, in my own experience, found of less size than one third of an inch in diameter, which, is a large size when compared with the minute specimens of *Calamodendron* and *Stigmaria*-like plants above alluded to.

It may be asserted with confidence that up to the present time we do not know of any specimen of *Lepidodendron* exceeding three inches in diameter, or any specimen of *Sigillaria* one foot in diameter, that affords evidence of its internal structure in anything like a perfect condition.

Assuming *Stigmaria* and *Halonia* to be roots, they have this marked difference, namely, the former, having attained its usual characters, is not found to bifurcate, whilst the latter appears to continue to bifurcate as far as it has been possible to trace it.

Great caution is required in attempting to connect the different, though allied, fossil plants by a series of gradations from one to another, taking a little from each, and then joining those parts together, more especially if the describer be an accomplished draughtsman; for his pencil is often too apt to be directed by his preconceived ideas rather than by the simple and true delineation of the specimen itself. I have endeavoured, as much as possible, to represent truly the appearance of the structures as seen in the specimens; and in order to do that I have had the advantage of securing the services of so correct an artist as Mr. J. N. Fitch, who gives exact delineations of the specimens as they appear to his unbiassed eyes, and not according to any preconceived opinions of my own. The endeavour has been to describe correctly the specimens without attempting to generalize, or to bind up into a whole scattered and fragmentary specimens, which may or may not have been formerly connected with different plants.

The opinions of various authors on the subject, up to the present time, are given in their own words ; but the facts observed by me are left to speak for or against such views, without specially combating or supporting them.

IV. Description of the Specimens.

§ 1. The Specimen (*Lepidodendron Harcourtii*), No. 31. Plate XIII (and No. 18, p. 48, Plate VII, fig. 6).

For this beautiful fossil (Pl. XIII, figs. 1—6) I am indebted to the kindness of my friend Mr. J. S. Dawes, who discovered it in the clay-ironstone of the Coal-measures near Dudley. The original specimen, however, from which the slices were made has unfortunately been lost or mislaid ; but, from what we can learn, we understand that it exhibited all the external characters of rhomboidal scars, and other appearances, which generally distinguish *L. Harcourtii*.

The transverse section of the specimen, magnified three and three quarters diameters, in fig. 1, is irregularly oval, measuring one and a half inches across its major, and one inch across its minor axis. The medulla is nearly circular, and measures two eighths of an inch in diameter. For about one tenth of an inch within the circumference of the specimen there appears, for the greater part of the distance, a line of division, as if there had been originally some change of structure there.

The central axis is eccentric, as is generally found to be the case with specimens of *Lepidodendron* affording evidence of the former existence of a medulla. It has been considered that such displacement has arisen from the destruction of the cellular tissue surrounding it ; but in this instance nearly the whole of the structure, from the centre to the circumference, has been beautifully preserved ; and therefore the displacement of the central axis in this instance can scarcely be attributed to that cause.

The celebrated Hesley Heath specimen, so fully illustrated and described by Witham, Lindley and Hutton, and Brongniart, and from which nearly the whole of our knowledge of the structure of *Lepidodendron* has been obtained, was by no means in a good state of preservation, so far as its medulla was concerned. This was assumed to consist of common parenchymatous tissue, like that constituting the greater portion of the stem exterior to the vascular cylinder. In the specimen now under description we shall find that the medulla, as represented on a large scale in Plate VII, fig. 6, of Part II of this Monograph, shows in a distinct manner every cell ; but these are not those of cellular tissue, such as is usually found to constitute ordinary pith, but large oblong cells, arranged in vertical series, exactly resembling those first described by Mr. Dawes as composing the medulla of *Halonia*. In my description of *Sigillaria vascularis*,[1] many years

---

[1] 'Quart. Journ. Geol. Soc. London,' for May, 1862, vol. xviii, p. 106, pl. iv.

ago, some of this structure was noticed in the medulla of that plant, intermixed with the scalariform tissue forming the bulk of the central axis; and a somewhat similar structure is mentioned as composing the medulla of *Calamodendron commune*, in Specimens Nos. 3 and 4, at page 23 of Part I of this Monograph. I have also observed it in the medulla of a small stem which would pass for a *Stigmaria;* and it has been there sometimes mistaken for common parenchymatous tissue. For the sake of convenience, and in order to distinguish it from ordinary parenchyma and prosenchyma, I purpose to term it " orthosenchyma," and " orthosenchymatous tissue," in the following descriptions.[1]

The medulla of orthosenchymatous tissue is surrounded by a zone of scalariform vascular tissue, of large dimensions in the interior, and gradually diminishing in size as it approaches and enters the dark corrugated line on the exterior. From this part originate the vascular bundles which communicate with the leaves. They appear to come from the concave spaces forming the outer edge of the dark line. Succeeding the vascular cylinder, which is quite destitute of any traces of medullary rays, is a dark shaded zone, of about twice the diameter of the internal woody cylinder previously described, composed chiefly of the mouths of the vascular bundles, but also exhibiting a series of round masses of large cells, of a light colour, which are only slightly shown in the plate (fig. 1). Next comes the great mass of fine parenchymatous tissue, traversed by vascular bundles; but this in the Hesley Heath and most other specimens has for the most part disappeared, having been replaced by mineral matter. This zone of tissue increases in size and strength as it approaches the broken and irregular line seen near to the exterior of the specimen. Outside this line the tissue gradually enlarges into long utricles, or tubes, arranged in a radiating series. This is covered by a thick epidermis, now converted into bright coal, showing no traces of structure.

Fig. 2 is a longitudinal section of a portion of the stem (magnified four diameters), extending from the inside of the woody cylinder (*b*), through the zone of fine vascular tissue (*c*), whence arise the vascular bundles (*d*), which traverse the mass of parenchymatous tissue (*e*) out to the leaves. These bundles at first run nearly parallel to the woody cylinder, then gradually curve outwards, until they become nearly level, when they again curve upwards, and reach the outside of the stem, and communicate with the leaves.

Fig. 3 is another longitudinal section of the inner portion of the stem (magnified seven diameters), showing the orthosenchymatous tissue (*a*) of a part of the medulla, the zones of coarse and fine vascular tissue (*b* and *c*), and the vascular bundles of scalariform tissue (*d*), surrounded by a zone of fine orthosenchymatous tissue similar in structure (but finer

---

[1] Probably some may object to the coining of a new term, and think that orthosenchyma is merely muriform tissue set upon its end; but that is scarcely so; for the short cross lines are not placed at right angles to the long ones, as is generally the case in the latter structure, but at different angles; and it cannot well be termed cubical cellular tissue.

in texture) to that of the medulla traversing the great mass of parenchyma (*e*), which forms the chief portion of the stem.

Fig. 4 shows a longitudinal section of an outer portion of the stem (magnified twelve diameters), in which are displayed the coarse parenchymatous tissue (*e*), passing into elongated utricles and tubes (*f*), that nearly resemble true vascular tissue, but show no markings on their walls, and the epidermis (*g*), converted into coal, about twice the thickness of the radiating zone.

Fig. 5 (magnified twenty-four diameters) is a partly transverse section of one of the vascular bundles, taken near the outside of the stem. In structure, whether we consider the vascular bundle of scalariform tubes, the orthosenchymatous tissue so often wanting, the light-coloured zone of fine parenchyma, and the outside of coarse parenchyma, it very much resembles a transverse section of a rootlet of *Stigmaria*, as figured and described by me in the 'Quart. Journ. Geol. Soc.,' vol. vi, pp. 18, 19. This structure had long previously been made out by Professor Goeppert.[1]

Fig. 6 (magnified five diameters) is a section of a portion of the outside of the stem, showing the coarse parenchymatous tissue, passing into the elongated utricles (or tubes), arranged in radiating series, together with a portion of the epidermis, and one of the outer bundles near to its passage into a leaf.

If we separate the various structures already described into two series, the one axial and the other epidermal, we have (1) an axis, composed of orthosenchymatous tissue, (2) a woody cylinder, formed of an inner zone of coarse vascular tissue and an outer one of finer vascular tissue, from the latter of which originate (3) the vascular bundles, passing to the leaves, belonging to the first, and (4) the zones of fine and coarse parenchyma, the latter passing into (5) elongated utricles or tubes, resembling those found in *Sigillaria vascularis*, *Stigmaria*, and *Halonia*, and (6) the outer bark, belonging to the second.

This specimen, to a great extent, confirms the views of Brongniart on the structure of the stem of *Lepidodendron Harcourtii*, as restored by him from the Hesley Heath specimen ; but the medulla is in a much better state of preservation than it was in the Rev. Mr. Harcourt's fossil. The vascular bundles are shown to be enveloped in a zone of orthosenchymatous tissue, which Brongniart, it appears, did not notice. The line of demarcation betwixt the inner and finer parenchyma and the coarser and outer zone is found to be nearly imperceptible, and not so well defined as he represented it. The outer radiating series, moreover, is composed of much longer utricles (or tubes) than those he restored. Altogether the characters of Mr. Dawes' specimen, though in a better state of preservation than that of Hesley Heath, do not, in my opinion, vary sufficiently from that plant to be made the basis of a new species ; therefore it has been considered desirable to class this specimen under *Lepidodendron Harcourtii*.

---

[1] 'Genera Plant. Foss.' (Stigmaria), p. 29.

Before leaving the subject of the structure of the stem of *Lepidodendron*, it will be as well very briefly to consider the dichotomization of the stem of this plant. This appears to have been of the same character in *Lepidodendron*, *Sigillaria vascularis*, *Halonia*, and probably other similar plants. It is like that which prevails in Lycopodiaceæ as described by Brongniart[1] as follows :—" Ce mode de ramification me paraît extrêmement rare parmi les plantes appartenant à d'autres classes du règne végétal, car toutes les plantes phanéro-games qui offrent des tiges dichotomes doivent cette apparence ou à un rameau réellement latéral et secondaire, qui a pris un accroissement égal au rameau principal, ou à deux rameaux latéraux, opposés ou alternes et rapprochés, qui se sont seuls developpés tandis que la tige principale s'est transformée en un simple pédoncle floral, ou bien a subi un avortement complet.

" Dans ces divers cas un des rameaux ou même tous les deux sont d'un ordre dif-férent de la tige à laquelle ils font suite, et ils naissent de l'aisselle d'une feuille insérée sur cette tige. Dans les Lycopodiacées, ou contraire, le développement est continu, et la tige tout entière se divise en deux faisceaux, comme on le voit quelquefois parmi les plantes phanérogames, dans les tiges monstrueuses, dites fasciédes, qui seules me paraissent offrir un mode de division analogue, malgré son irrégularité, à celui des Lycopodiacées." This mode of division in the stem has not to my knowledge been hitherto noticed, although collectors must have observed instances of it.

§ 2. The Specimen (*Lepidodendron Harcourtii*), No. 32. Plate XIV, figs. 1, 2, 3.

Specimen No. 32 (Plate XIV, fig. 1, natural size) is the transverse section of a stem of *Lepidodendron Harcourtii*, found by me in the Black-shale-ironstone in the lower part of the Middle Coal-measures at Hady, near Chesterfield. It is of an oval form, and measures three inches across its major and one and a half inches across its minor axis. The outside is marked with prominent rhomboidal scales, measuring three eighths of an inch one way and two eighths of an inch the other. The fossil, composed of clay-ironstone of a black colour, is enveloped in a coating of bright coal, and shows no trace of structure except in the two horseshoe-shaped bodies in the inside. At the first glance these might have been considered to have been accidentally introduced after the interior of the stem had been decomposed and removed ; but the occurrence of similar axes in the stems of *Sigillaria vascularis* and *Halonia regularis*, hereinafter described, shows that this is very improbable, if not impossible. The two axial bodies occur in about the middle of the specimen, and at about equal distances from the sides, each measuring two eighths of an

[1] 'Histoire des Végétaux fossiles,' Tome ii, pp. 3, 4.

inch across. There is no doubt that we have here the stem of a *Lepidodendron* cut across where the woody cylinder has commenced to divide, but lower down than where the whole stem dichotomizes into two, as is so generally seen in *Lycopodiaceæ*.

Figs. 2 and 3 (magnified three and a half diameters) represent the two vascular axes. Their outer portions show structure, in the large scalariform tubes generally found constituting the woody cylinder of *Lepidodendron*. There is no trace of anything resembling medullary rays. The whole of the medulla has been replaced by white carbonate of lime. Fig. 2 represents the lower and fig. 3 the upper of the bodies seen in fig. 1.

§ 3. THE SPECIMEN (*Sigillaria vascularis*) No. 33. Plate XIV, figs. 4, 5, 6.

Specimen No. 33 (Plate XIV, fig. 4, magnified three and a half diameters) is a transverse section of *Sigillaria vascularis*, found by me in the "Bullion" seam of coal (marked ** in the section of the Lancashire Coal-measures, p. 12) at Spa Clough near Burnley. It is of an elongate-oval form, and measures one and six eighths of an inch across its major, and seven eighths of an inch across its minor axis. The specimen is enclosed in its matrix of carbonate of lime. The two apple-shaped vascular axes, parted by some fine orthosenchymatous tissue, and surrounded by a radiating cylinder of vascular tissue of a wedge-shape, containing medullary rays, appear to have been enclosed in a zone of delicate parenchyma, which, having partially disappeared, has been more or less replaced by carbonate of lime. This is succeeded by a zone of strong parenchyma, which gradually passes into long utricles or tubes, arranged in radiating series; and throughout all pass the vascular bundles, leading from near the central cylinder to the leaves. The external covering appears to have consisted of an outer bark, now converted into coal.

Figs. 5 and 6 (magnified ten diameters) represent the right and left apple-shaped bodies free from the stem. Their axes are seen to be composed of large scalariform utricles or tubes, and of fine orthosenchymatous tissue near the notches in the outer lines. They are surrounded by a cylinder of radiating vascular tissue, and a portion of the zone of fine parenchyma envelopes it.

A similar example of dichotomization of the stem of *Halonia regularis*, hereinafter described, also shows the division of the vascular axis of the stem into two, like those of *Lepidodendron* and *Sigillaria*.

Brongniart, in speaking of the stems of *Lycopodiaceæ*,[1] says, " Parmi ces dernières [Lycopodiacées] ce mode de ramification tient à leur développement entièrement terminal,

[1] 'Histoire des Végétaux fossiles,' Tome ii, p. 4.

mode de développement que cette famille partage avec les Fougères, et probablement avec quelques autres familles voisines, et qui devrait faire réserver à ce groupe de végétaux le nom très-juste d'Acrogènes, appliqué par M. Lindley à toutes les Cryptogames et Agames."

This no doubt is true with regard to *Lycopodiaceæ;* but with respect to *Lepidodendron, Halonia,* and *Sigillaria,* the two first-named genera exhibiting no traces of medullary rays, whilst the latter possesses them, are very like Acrogens; but most assuredly the last-named genus, and probably the other two, were exogenous plants, as the specimens formerly described by me distinctly prove.[1]

V. Bibliographical History of Halonia.

Having thus concluded my description of the internal structure of the stem of *Lepidodendron* and its mode of dichotomization, I will now proceed to describe some specimens of *Halonia regularis.* First, however, it is desirable to give a short sketch of the present state of our knowledge of this fossil plant.

Several authors have distinctly stated that *Lepidodendron* had roots resembling those of *Stigmaria;* and Mr. Dawes, who possessed in 1848 by far the most perfect specimens then known, hinted that *Halonia* might prove to be the root of *Lepidodendron.*

§ 1. *Lindley and Hutton.*

a. " Halonia? tortuosa.[2]—In sandstone in a quarry near South Shields, from a specimen furnished by Isaac Cookson, Esq.

" Whatever this may have been, it is evidently very distinct from anything hitherto described. Probably the present specimen has been jammed and distorted so much as to have lost, in a great measure, its original character; but enough remains to convey some idea of its external structure.

" It seems to have been a plant of small dimensions, the surface of whose stem was completely covered with little processes, which, in falling away, left minute quincuncial ill-defined spots, that rapidly became separated and obliterated as the stem advanced in age. Among these spots, at intervals of three fourths of an inch every way, were arranged little projections, the apex of which was terminated by some appendage now lost. The ramification appears to have been dichotomous, but this is extremely uncertain.

---

[1] ' Quart. Journ. Geol. Soc.' for May, 1862, vol. xviii; and ' Phil. Trans.,' volume for 1865, p. 576.
[2] Lindley and Hutton's ' Fossil Flora,' vol. ii, p. 85, 1833—5.

"The principal questions to answer are, first—What were the processes? And, second, What were the projections? If the processes were leaves, as it appears probable, then the projections will have been either the bases of old or the points of rudimentary branches; and in that case the affinity of the fossil will be nearest with *Halonia*. (See the next article.) But if we suppose the processes to have been analogous to the ramenta of Ferns, then the projections may be considered to be of the same nature as those we find in *Stigmaria*, where they are plainly the bases of leaves. A great objection to this view is, that the arrangement of the spots left by the processes is too regular for ramenta.

"The only branch that is seen in the specimen will not enable a botanist to say whether the mode of ramification was dichotomous or alternate. If the projections are the bases of leaves, it may have been dichotomous; but, if they are rudimentary branches, it must have been alternate.

"Under these cirumstances we are forced to leave the specimen in a state of uncertainty, which is unfortunately too common in this science."

b. "*Halonia gracilis* (page 86).—From the Coal-measures of Low Moor in Yorkshire.

"At first sight one would be disposed to consider this a *Lepidodendron*, to which its rhomboidal scars give it strong resemblance. But if we consider *Lepidodendron* as an extinct form of *Lycopodiaceæ*, we must limit it to those fossils in which the mode of branching was dichotomous, for no other kind of ramification is met with in recent *Lycopodiaceæ*.

"Here, however, it is plain, from numerous scars of branches, that they were arranged in an alternate manner round a common elongating axis, after the plan as now obtains in the Spruce Fir. In fact, if we compare this to a vigorous branch of a Spruce Fir, one year old, we shall find the resemblance very striking even in the scars of the leaves. For this reason, and for the sake of rendering our notions of the extinct *Flora* as definite as we can, the genus *Halonia* is proposed, to comprehend all those fossils in which the surface of *Lepidodendron* is added, with the mode of branching of certain *Coniferæ*, and which, it is therefore to be inferred, were of a nature analogous to the latter."

c. "*Halonia regularis* (vol. iii, p. 228).—Fig. 1, from Halliwell Stone Quarry, near Bolton. Fig. 2, from Peel Stone Quarry, near Bolton, both communicated by Mr. Dawes, of Bolton.

"These are most remarkable specimens of this curious genus. They are quite distinct, both in dimensions and in the regularity with which their tubercles are arranged, from either of the species previously figured."

§ 2. Mr. RICHARD BROWN, in a description of an upright *Lepidodendron* with *Stigmaria* roots, in the roof of the Sydney Main-coal, in the Island of Cape Breton, states,[1] "Since I forwarded to the Society a description of the Sydney *Sigillaria*, about twelve

---

[1] 'Quart. Journ. Geol. Soc. of London,' vol. iv, p. 46, 1847.

months ago, I have discovered several upright trees in the coal-measures, evidently *not Sigillariæ*, with roots of *Stigmaria* united to them. These trees exhibited so many of the characters of *Lepidodendron* that I at once concluded they belonged to that genus; but, having never seen it hinted that *Lepidodendron* possessed *Stigmaria* roots, and distrusting my own skill in fossil botany, I determined to wait till I could collect more decisive evidence in confirmation of my opinion. This evidence I have now obtained in another example, fortunately most complete in all its parts, a description of which I hasten to lay before the Society, accompanied with sketches, which I hope will clearly prove that the stem in question is a genuine *Lepidodendron* united to roots of *Stigmaria*."

§ 3. UNGER, in 1847, in his 'Chloris Protogæa,' under his Order XX, *Lepidodendreæ*, 98, *Halonia*, Lindl. and Hutt. says:

"Trunci arborei cylindrici decorticati? cicatricibus minoribus punctiformibus vel rhomboideis spiraliter dispositis, majoribus tuberculatis remotis instructi."

"Lindl. and Hutt., 'Foss. Flora,' ii, p. 12.

"1. *Halonia tuberculata*, 'Brong. Hist. Végét. Foss.,' ii, pl. xxviii, figs. 1, 2, 3.

"*Halonia tortuosa*, Lindl. and Hutt., Foss. Flo., South Shields, vol. ii, pl. lxxxv, p. 11.

"2. *Halonia gracilis*, Lindl. and Hutt., Low Moor.

"3. *Halonia Beinertiana*, Göpp., p. 203, Charlottenbrun, Silesia."

§ 4. Mr. J. S. DAWES, in some remarks on the internal structure of *Halonia*,[1] states that "it was proposed by the authors of the 'Fossil Flora' that the genus *Halonia* should comprehend all those plants combining the surface of the *Lepidodendra* with the mode of branching of the *Coniferæ*, to which latter order they considered these fossils to be analogous. The discovery, however, of better preserved specimens has clearly shown that the supposed remains of alternate branches, noticed more particularly in the species *H. gracilis*, must have been merely impressions of the protuberances which characterize these fossils, and that they are, in fact, like the *Lepidodendra*, dichotomous. A still further proof of cryptogamic affinity is now afforded by sections of a specimen from the neighbourhood of Birmingham, in which traces of the vegetable structure have been preserved. By reference to the drawing, fig. 1, it will be seen that this stem is composed of a central medullary column (*a*), surrounded by a series of scalariform vessels (*b*); these being succeeded by a compact cellular tissue (*c*), which becomes more lax between this central part and the cortical zone, the latter (*d*) being composed of a thick-membraned, very regular tissue, and bearing a large proportion to the rest of the stem, equal in some specimens to one third of the diameter. There are no concentric rings, or, strictly speaking, medullary rays, neither any ligneous fibre, or indeed any indications whatever of affinity with the *Coniferæ*, or even with that division of the Dicotyledons, except that some

---

[1] Vol. iv, pp. 289—291, of the 'Quart. Journ. Geol. Soc.,' 1848.

similarity exists in the character of the striated tubes which surround the medullary column, and the pseudo-vascular bundles of certain *Zamiæ*. Neither are these plants to be referred to that class which includes the *Sigillaria, Anabathra*, &c.; for, although the structure in some important respects may correspond, the arrangement of the tubes of the vascular system is altogether reversed; consequently, the curved scalariform bundles, which traverse the stem from its axis to the periphery, do not emerge from the tissues immediately in connection with the medullary column, but are thrown off from the outer portion of the sheath. These leaf-cords, which appear somewhat to resemble the stem in miniature, take a direction for some distance nearly horizontal, so that different portions of the tissues of several neighbouring bundles are usually cut through, giving to the transverse section some appearance of a radiated structure. I should observe, that these fasciculi differ in size, the smaller ones having a direction towards the spirally arranged scars which cover the surface of the stem; the larger ones being connected with the processes that occur upon it at certain intervals, each of these projections exhibiting a roundish cicatrix at its apex, as though some leaf-like appendage had been supported upon it, and having some resemblance to the well-known tubercles of *Stigmaria*. These few observations will be sufficient to show that the fossil in question belonged to the vascular *Cryptogamiæ*; and that, when compared with the other plants of the Coal-measures, the nearest affinity is with *Lepidodendron*. We might, in fact, considering their tortuous root-like appearance, and on other accounts, be tempted to speculate as to the relationship they bear to this fossil; but possibly some other specimens in my possession, not yet sufficiently examined, may throw further light upon the subject.

"Since the above remarks were forwarded to the Society, I have been fortunate enough to obtain some very good sections of another specimen of this fossil, and am now enabled to mention a peculiarity in the structure which had previously escaped notice, viz., that a narrow ring of very regular, compact, elongated tissue exists on the outer portion of the cortical zone (*e*), similar to the prosenchymatous arrangement mentioned as occurring in the corresponding part of the *Lepidodendron*. Having, however, had an opportunity to look through many specimens of this latter fossil, I may venture to say that the descriptions hitherto given of it do not in this and some other respects correctly represent its structure. Such discrepancies have probably arisen from the inferior state of the specimen first met with by the Rev. C. G. Vernon Harcourt, and also in consequence of Mr. Witham having originally figured from portions of two distinct fossils, apparently mistaking in one instance an imperfect fragment of *Halonia* for a piece of *Lepidodendron* (see 'Transactions of the Natural History Society of Newcastle,' 1832, and 'Internal Structure of Fossil Vegetables,' Edinburgh, 1833, pl. xii, fig. 3; pl. xiii, fig. 1). Brongniart indeed admits being unable to detect this exterior tissue; but nevertheless describes it both in his 'Histoire des Végétaux Fossiles' and in the 'Archives du Muséum d'Histoire Nat"relle' upon English authority; he has, however, discovered a very similar tissue, although differently placed, in the cortical zone of the *Sigillaria elegans*.

"There are some other points connected with, and in the constitution of, these fossils that I hope to refer to on a future occasion; and may perhaps now observe that the medullary column does not, either in the *Lepidodendron* or *Halonia*, consist of the usual parenchymatous tissue, but seems to be composed of large quadrangular cells arranged in perpendicular series, and presenting an appearance as though each minute column was confined within a slight membrane or tube. I believe that no such structure has been found to exist in recent vegetation, the nearest approach to it being probably in *Psilotum*, one of the Lycopod family, and of course incompatible with the idea of this central portion being a true medulla; these plants must therefore be still further removed from any supposed phanerogamic alliances."

§ 5. Dr. HOOKER, in writing of *Lepidodendron*,[1] states, "Of the stem, branches, leaves, and fructification, we have thus a satisfactory knowledge, but the nature of their roots is not ascertained. Mr. Dawes, of West Bromwich, to whom I am indebted for much information regarding the structural character of coal fossils, is inclined to regard the species of *Halonia* as roots of *Lepidodendron*, on which opinion I have no remarks to offer." Again, on the same page, he observes, "*Halonia*, another genus of *Lepidodendreæ*, is composed of three species, possibly the roots of these or others of the same genus."

§ 6. BRONGNIART, in treating of *Halonia*, writes,[2]

"HALONIA, *Lind. et Hutt.*

"Les tiges, assez rares et mal connues, qui forment ce genre, offrent, sur les parties qui sont bien conservées, une écorce, marquée de cicatrices foliaires, disposées comme dans les *Lepidodendron*; mais la tige présente en outre de gros tubercules coniques disposés en quinconce, et sur lesquels s'étend uniformémént l'écorce générale et les feuilles qu'elle supportait.

"La disposition quinconciale des mamelons ou tubercules qui font saillie sur la tige et la continuitié de leur base avec le reste de l'écorce de la tige distingue complètement ce genre des précédents. Ici les gros mamelons ne paraissent pas des cicatrices, mais des saillies sous-corticales comme celles qui seraient produites par des racines non sorties de dessous l'écorce."[2]

---

[1] "On the Vegetation of the Carboniferous Period," 'Memoirs of the Geological Survey of Great Britain,' vol. ii, Part II, p. 422, 1848.

[2] 'Tableau des Genres de Végétaux Fossiles,' p. 43, 1849.

At page 96 he gives the following classification :
" LYCOPODIACÉES.
Lépidodendrées.
Lepidodendron.
Lepidostrobus.
Lepidophyllum.
Ulodendron.
Megaphyton.
Halonia.
Lepidophloios.
Knorria."

§ 7. GOLDENBERG, in writing on *Lepidodendron*, appears to consider that *Halonia* is a genus allied to that plant, if not even a species of the same genus.[1]

VIII. *Genus* HALONIA.

" LEPIDODENDRON.—The stem of these plants is cylindrical ; and on the bark of well-preserved specimens the points of attachment of the leaves may be seen, which resemble in all points those of *Lepidodendron*. But in addition to these the stem also bears conical protuberances, arranged quincuncially, which are evenly covered by the bark with its leaf-scars, so that it seems as if these protuberances were due to branches not yet come through.

" If the form and dimensions of the parts where the leaves were attached suggest that some species of *Lepidodendron* is represented by *Halonia*, it is still more confirmed by our often having met with specimens of *Halonia* exhibiting the same ramification as *Lepidodendron*.

" The specimen represented and described by us shows that the protuberances appeared on the outer branches only, there being no traces of such swellings below the bifurcation ; and, lastly, in favour of this view is the fact that all the stems of *Haloniæ* have a proportionally small circumference.

" 39. *Halonia dichotoma*, pl. iii, fig. 12.

" This three-inch stem is branched dichotomously, and bears spirally arranged papillæ on each arm of the bifurcation, whereas the stem never bears these growths below the bifurcation.

" The bark is covered with rhomboidal leaf-scars, resembling in shape and arrangement the leaf-attachments of *Lepidodendron*.

[1] ' Flora Saraepontana fossilis,' part i, p. 20, 1855. For this translation I am indebted to the kindness of my friend, Mr. Charles Bailey, the indefatigable Librarian of the Literary and Philosophical Society of Manchester.

"Our specimen was found in the railway-cutting at Friedrichsthal. Here also are found

"40. *Halonia tuberculata*, Brong., 'Hist. Vég. fos.,' ii, p. 28, figs. 1, 2, 3. Auerswaldflotz, near Gerweiler.

"41. *Halonia regularis*, Lind. and Hutt., 'Foss. Flor.,' iii, p. 179, pl. 228. From the Carboniferous Sandstone at Duttweile.

"§ 8. *Binney.*—In a paper of my own on *Sigillaria* and its roots,[1] a full account is given of specimens observed, and the date of their publication. It is there stated, "My specimens here mentioned beyond doubt prove that *Stigmaria* is the root of *Sigillaria*. It will probably be found that this character of root was common to several other genera of aquatic plants besides *Sigillaria*, but this point requires carefully to be examined and proved."

§ 9. Dr. SCHIMPER describes *Halonia* in the following words :[2]—"Trunci dichotome ramosi, ramis patentibus, mediocriter crassi, corticati rhombeo-cicatricosi, tuberculati; decorticati breviter quincunciatim papillosi, tuberculis apice perforatis basi vel totâ superficie papillis brevibus (pulvinulis foliorum subcorticalibus) tectis, spiraliter dispositis. Cylindrus centralis medulla impletus.

"Type végétal très-bizarre, dont le mode de ramification, la forme et la disposition des cicatrices foliaires rappellent bien les *Lepidodendron*, mais qui se distingue de ce genre par un système de tubercules obtus-coniques disposés en quinconce, qui recouvre la tige et dont la signification morphologique n'a point encore été déterminée. M. d'Eichwald y voit les points d'attache des feuilles, et dans les petites cicatrices rhomboidales celles d'écailles qui auraient recouvert la tige. Cette manière de voir est évidemment erronée. M. Goldenberg croit voir dans ces bourrelets, qui, dans les échantillons bien conservés, sont couverts de cicatrices foliaires comme le reste du tronc ou du rameau, des rameaux à l'état latent non arrivés à leur développement normal. Si ces proéminences sont percées d'une ouverture vasculaire à leur sommet, comme M. d'Eichwald les représente (voy. notre planche) il est naturel d'admettre, ce me semble, que c'étaient là les points d'attache des fruits.

"Le genre *Halonia* est limité au terrain houiller, où il est représenté par un très petit nombre d'espèces, si toutefois il y en a plusieurs, et par peu d'individus.

"3. *Halonia regularis*, Lindl. et Hutt. Tuberculis 6-seriatis foliorum cicatricibus (vasorum cicatriculis) subcorticalibus punctiformibus. Lind. et Hutt., *op. c.* iii, p. 179, pl. 208.

[1] 'Transactions of the Manchester Geological Society,' vol. iii, p. 110, May, 1861.
[2] 'Traité de Paléontologie végétale, ou la Flore du monde primitif,' &c. Tome seconde, Première Partie, p. 53.

"Dans le grès houiller de Halliwell et Peel, près de Bolton (Angleterre), près de Duttweiler, pays de Saarbrucken."

Dr. Schimper, in a supplementary note of the same work, at p. 117, states that, after his general considerations on *Stigmaria* had been printed off, he heard of the discovery (in the quarries of the grauwacke, near Thun) of a trunk furnished with roots. This trunk has been obtained by Doctor Faudel for the Museum of Colmar; and when he (Dr. Schimper) examined the specimen, it completely confirmed his opinion previously expressed that *Stigmaria* might have belonged to other trees besides *Sigillaria*. He found, in fact, that the trunk in question furnished with a Stigmarioid root was indeed a trunk of *Knorria longifolia*, whose base agreed with *Ancistrophyllum*, while its middle portion corresponded with *Didymophyllum Schottini*, Gœpp.

VI. Description of the Specimens.

§ 1. Specimen No. 34, *Halonia regularis*. Pl. XV, figs. 1—4.

For this instructive specimen (Pl. XV, fig. 1, magnified three and one fifth diameters) I am also indebted to Mr. Dawes, who discovered it in the clay-ironstone of the Coal-measures near Dudley. The original fossil from which the slices were made is not now to be found, so I cannot describe, from my own knowledge, what were its external characters; but, from Mr. Dawes' description, it no doubt belonged to the genus *Halonia*; and, from its identity in structure with other specimens of my own, intended to be hereinafter described, it most probably belonged to the species *regularis*.

The transverse section of the specimen, as shown in fig. 1, is irregularly oval, flattened on one side, and measures one and five eighths inches across its major, and one and three eighths inches across its minor axis. The central axis is near the middle of the specimen. The medulla in this section appears to be much disarranged and destroyed, so that we cannot tell whether it consisted of orthosenchymatous tissue, like that in *Lepidodendron*, or not. The woody cylinder, enclosing the medulla, is also considerably disarranged; but there is sufficient evidence to show that it was composed of barred tubes or utricles. The mouths of some of these appear to be filled with a very fine orthosenchymatous tissue, of a bright wine colour, probably a portion of the medulla squeezed into them during the process of mineralization. The outside of the woody cylinder is bordered by an irregular band, dark in colour, as drawn in the plate; but, when viewed by transmitted light, it is seen to be of a bright wine colour. From near this zone proceed the large vascular bundles, which go outwards through the parenchyma to the leaves or roots (whichever of those appendages they may prove to be). The portion of the stem from the woody cylinder to the irregular light-coloured line near the outside of the specimen consists of fine parenchyma, on which are dotted numerous mouths of vascular bundles, enveloped in orthosenchyma, similar to those found in *Lepidodendron* previously described. Outside of the above-named line, where the parenchyma appears stronger, and its cell-walls thicker, it seems gradually to pass into elongated utricles (or tubes), and to assume at the edge a radiating appearance. In the specimen there appears to be only a little coaly matter near the upper part on the right hand side of the finger, the other portion probably having been removed. The transverse section, on the whole, is like that of *Lepidodendron;* and the great value of the specimen is the perfect preservation of the delicate parenchymatous tissue surrounding the woody cylinder, and so generally absent in woods of this description. It supplies a part wanting in a valuable specimen (No. 35), to be hereinafter described.

Fig. 2 (magnified four diameters) represents a longitudinal section of a portion of the same specimen, showing the medulla (*a*), composed of orthosenchymatous tissue, in a beautiful state of preservation; the woody cylinder, formed of scalariform utricles or tubes (*b*), the fine scalariform tubes or utricles (*c*), from which proceed the large vascular bundles (*d*), composed of scalariform tubes, enveloped in a zone of fine orthosenchymatous tissue, traversing the parenchyma (*e*), and leading to the rootlets. Two of these bundles are seen going down from the woody cylinder on the left hand side, and one on the right hand side. The woody cylinder has been displaced, in this specimen, from its original position, and now appears sloping instead of vertical. On the extreme edge of the fossil there is evidence of some traces of the elongated utricles or tubes (*f*), the occurrence of which has been previously alluded to in the description of the transverse section.

Fig. 3 (magnified eight diameters) exhibits a longitudinal section of a portion of the medulla (*a*), the woody cylinder (*b*), and the darker zone (*c*), whence the vascular bundles (*d*) appear to originate, all in a most beautiful state of preservation.

Fig. 4 (magnified twenty-five diameters) gives a transverse section (nearly at a right angle) of one of the vascular bundles, enveloped in a zone of orthosenchyma near the outside of the specimen. This, in all its parts, is scarcely to be distinguished from the vascular bundle taken from the same part of *Lepidodendron*, and hereinbefore described and figured in page 79, Pl. XIII, fig. 5, or from a similar vascular bundle in *Stigmaria*.

§ 2. Specimen No. 35, *Halonia regularis*. Plate XVI, figs. 1—5.

Specimen No. 35 (Pl. XVI, fig. 1, natural size) represents an example of *Halonia regularis*, from the Upper Brooksbottom seam of coal in Lancashire, marked * in the section of strata previously given at page 12 of this Monograph. It is imbedded in a matrix of limestone, and has only a part of one side exposed. The specimen is one and three eighths of an inch long; and its section is irregularly oval, measuring one and one eighth of an inch across the major, and six eighths of an inch across the minor axis.   The outside shows numerous small superficial papillæ, and four circular tubercles, each having a depressed areola, with a small projection in the centre.   The medulla and vascular cylinder are in a most perfect state of preservation ; but the zone of fine parenchymatous tissue, so well preserved in Specimen No. 34, is for the most part wanting, having been replaced by carbonate of lime.   In this mineralized portion is seen a vascular bundle, showing structure.   Next comes a zone of parenchyma, increasing in the strength of its cells until it gradually passes into elongated utricles (or tubes), arranged in radiating series, constituting the outside next the bark.

Fig. 2 (magnified five diameters) represents a transverse section of the specimen, showing the medulla composed of orthosenchyma, and a woody cylinder, formed of large scalariform utricles (or tubes) in the interior of the circle, and smaller ones next the outside, ftom which proceed the vascular bundles.   One of the last is seen considerably distorted in appearance, but showing the structure in the central part.   The thick zone of parenchyma gradually increases in strength, until it passes into the outer zone of elongated utricles or tubes.   It may be objected that the vascular bundle has been introduced into the interior of the stem after the decomposition of the delicate parenchyma surrounding the woody cylinder ; but this argument, if good for anything, would prove that the latter had also been introduced.   The position of this woody cylinder, however, and of the vascular bundle, occupying their original places, is clearly shown in the specimen, No. 34, last described, where those parts are seen imbedded in the parenchyma as when growing. No doubt the structure of this vascular bundle is much the same as that found in *Stigmaria*, and might easily be mistaken for it ; but it seems more reasonable to suppose that this vascular bundle belongs to the stem in which it is now found than to another that is not even seen near it.   In other specimens in my cabinet there is abundant evidence to prove that *Halonia regularis* was furnished with vascular bundles proceeding from the woody cylinder to the outside, in all respects like those found in *Lepidodendron* and *Stigmaria*, and previously described in this Monograph.

Fig. 3 is a longitudinal section of a portion of the whole root (magnified ten diameters), showing the medulla (*a*), composed of orthosenchymatous tissue, the large scalariform tubes forming the inside of the woody cylinder (*b*), and the smaller ones, composing the exterior (*c*), the fine parenchyma (*e*), and the elongated utricles or tubes, forming the outer radiating cylinder (*f*).   This section does not very distinctly show any vascular bundles proceeding from the woody cylinder, except a trace of one on the right hand side in the specimen ; but this is not more than might have been expected, for the section is only one quarter of an inch in length, and the tubercles or rootlets on the outside of *Halonia* are very few in number, when compared with the leaves on *Lepidodendron* and the vascular bundles seen in that plant.

Fig. 4 (magnified ten diameters) is a transverse section of the medulla, composed of orthosenchyma, surrounded by a woody cylinder of scalariform tubes or utricles, all in a most beautiful state of preservation, with scarcely a cell wanting, and nearly as perfect as when they grew.

Fig. 5 (magnified forty diameters) represents a nearly right-angled transverse section of the vascular bundle in the middle of the root. This we take for one of those which proceeded from the outside of the woody cylinder to the rootlets. It consists of an axis of scalariform tissue, composed of two different sizes of tubes, the one being much larger than the other, and formerly imbedded in a zone of orthosenchyma, which has disappeared in the specimen. This is succeeded by a zone of regular parenchyma.

On comparing this section with a similar one of *Stigmaria*, figured and described by me many years since, scarcely any difference can be detected:[1]

§ 3. SPECIMEN No. 36, *Halonia regularis*.   Plate XVII, figs. 1, 2, 3.

Specimen No. 36 (Plate XVII, fig. 1, natural size) is another example of *Halonia regularis* from the Upper Brooksbottom seam of coal. It is imbedded in a nodule of limestone, and only shows a portion of one side of the fossil. Its transverse section is irregularly oval, measuring one and a half inches across its major and three quarters of an inch across its minor axis. The length of the specimen is two and one quarter inches. So far as it is exposed, no one would be able to distinguish it from a small specimen of *Stigmaria ficoides*. Unlike No. 35, it presents no appearance of tubercles ; but in their places are depressed areolæ of a circular form, having a raised ball in the centre, and arranged in quincuncial order. From its external characters the most experienced judge of Carboniferous fossils would not be able to distinguish it from a *Stigmaria ;* but when we proceed to examine its internal structure, we find that it is exactly the same as that of *Halonia regularis*, and very different from that of *Stigmaria ficoides*.

Fig. 2 (magnified five diameters) represents a transverse section of the same specimen, which displays a medulla composed of orthosenchyma and a woody cylinder, formed of large scalariform tubes in the interior of the circle, and smaller ones next the inside, whence proceed the vascular bundles ; a trace of one of these is seen. A portion of the delicate parenchyma, next to the woody cylinder, has been destroyed, and replaced with mineral matter ; but the greater part of it is preserved, and shows it to increase in strength as it approaches the circumference, assuming a prosenchymatous structure and a radiating appearance. Outside of this is a dark line, succeeded by an epidermal layer whose structure is apparently destroyed. It is upon the surface of the latter, however, that the Stigmarioid areolæ occur ; and, if it were carefully removed, we should most probably find the tubercles generally seen, as in specimens Nos. 35 and 38, and other common forms of *Halonia regularis*, in the place of the areolæ in this specimen.

[1] 'Quart. Journ. Geolog. Soc.,' vol. vi, pp. 18, 19.

HALONIA.

Fig. 3 (magnified twenty diameters) represents the centre of one of the areolæ, showing the vascular bundles in the middle, with a zone of orthosenchyma, which has for the most part disappeared, as is generally the case, and a regular zone of parenchyma very similar to that exhibited in fig. 5 of Plate XVI, as occurring in specimen No. 35.

§ 4. Specimen No. 37, *Halonia regularis*. Plate XVII, figs. 4, 5, 6.

Specimen No. 37 (fig. 4, natural size) is another example of *Halonia regularis*, also from the Upper Brooksbottom seam of coal. Like the last described specimen, it is partly imbedded in a matrix of limestone. The portion exposed shows the main stem, or rather *root*, immediately before it dichotomizes, and parts of the forking branches. The transverse section of the former is irregularly oval, measuring one eighth of an inch across its major and rather more than half an inch across its minor axis. The two portions of the fork are nearly circular, and each measures about five eighths of an inch in diameter. The depressed areolæ, so well shown in specimen No. 36, are not equally displayed in this, but there is sufficient evidence to prove that it belongs to *Halonia regularis*.

Fig. 5 (magnified five diameters) is a transverse section of the specimen, showing the medulla composed of orthosenchyma, and a woody cylinder composed of large scalariform tubes in the interior of the broken circle, and lesser utricles or cells next the exterior. The division of the woody cylinder might at first sight have been taken for an accidental disruption before the specimen was fossilized; but, as it occurs just before the root dichotomizes, and is not the only example of the kind in my possession, I am led to believe that it is the first commencement of the division of the woody cylinder into two, as previously described at length in specimens No. 32 and No. 33. Outside the woody cylinder the delicate parenchyma has, for the most part, disappeared; but evidence of four vascular bundles, partly showing structure, is seen in the mineralized portion; and, where the parenchyma becomes stronger and passes into prosenchyma, indications of more bundles are seen.

The exterior presents a thick epidermis, similar to that described in the last specimen (No. 36); and the two are so much alike in structure that it is impossible to distinguish the one from the other.

I am aware that the vascular bundles last named, as well as those in Nos. 35 and 36, have been generally described as intruding rootlets of *Stigmaria* which have penetrated the stem. No doubt they do appear larger, especially on their outsides, than we should expect to find in such a root; but the delicate tissues of those parts have nearly always been much distended, whilst the vascular bundles in the centre remain generally of the same size. Whether they be taken for the vascular bundles of the root, in which they are now found, or intruding rootlets of *Stigmaria*, their structure is no doubt the same, and therefore we cannot absolutely decide the question; but still it appears to me that it is under the circumstances more reasonable to believe that they are vascular bundles belonging to the body in which they are now found than similar organs belonging to another plant.

Fig. 6 (magnified 10 diameters) represents the woody axis, exhibiting structure throughout, except in the unshaded portion in the figure, which is composed of crystalline carbonate of lime, and which appears as if the medulla and woody cylinder had been disrupted prior to the mineralization of the specimen, and would no doubt have been taken as such did it not occur immediately before the dichotomizing of the root, and had not specimens Nos. 32 and 33, as well as others in our cabinet, been brought to our notice.

§ 5. Specimen No. 38 (*Halonia regularis*). Plate XVIII.

This is another specimen of *Halonia regularis*. It was found by Mr. Higson, one of Her Majesty's Inspectors of Coal-mines, and presented by him to the Manchester Museum. For permission to figure it I am indebted to the kindness of the Commissioners of that Institution. It is from the sandstone quarry of Peel, near Bolton-le-Moors, a locality noted for the beautiful specimens of Coal Plants which have been found there. The specimen shows no traces of internal structure, being simply a cast in a very fine-grained sandstone, with its thick epidermis converted into coal. The tubercles project from the stem about a quarter of an inch, and show how the spaces intervening between them had been once covered by a similar thick mass of coal. Where the outside matrix of the coal has been removed, the tubercles are seen covered by the coaly envelope, and only exposing a depressed areola, like that shown in specimen No. 36, and the areolæ so generally seen on the outside of *Stigmaria*. In fact, when seen with its epidermis preserved, it is a *Stigmaria* in outside appearance; but without that part of the plant it is a *Halonia regularis*. The exterior of the root, when deprived of its coaly envelope, is seen to be covered with numerous papillæ as well as the tubercles previously described. The length of the fossil is $9\frac{1}{2}$ inches, and its diameter 2 inches, and its form is nearly as cylindrical as it originally grew, apparently having been subjected to little or no pressure.

In specimen No. 34 the structure of *Halonia regularis* is beautifully shown, but nothing of its external character. In No. 35 the internal structure and the tubercles are given. In specimen No. 36 are exhibited the internal structure and the areolæ. In specimen No. 37 we have evidence of the dichotomization of the plant, and its internal structure. Lastly, in No. 38, although it affords no evidence of internal structure, there is a most perfect representation of all the external characters of the plant, without its being compressed, and before it was mineralized. The conditions in which these fossils often occur clearly show that when deprived of its epidermis the plant has generally been taken for *Halonia regularis;* but, when clothed with its epidermis, it clearly resembled, and might be taken for, a *Stigmaria.*

VII. CONCLUDING REMARKS.

Those portions of the Hesley Heath specimen in which the structure of the stem had been destroyed have fortunately been preserved in the Dudley specimen described in this Monograph, so that it may be now considered that we possess a knowledge of the structure of the whole of the stem of *Lepidodendron Harcourtii.*

In addition to the reasons given at page 37 for believing that *Sigillaria* and *Lepidodendron* were different, though allied, plants, I consider that the Dudley specimen of *Lepidodendron Harcourtii* clearly proves that it had a medulla of orthosenchyma, whilst the *Sigillaria vascularis* and *Lepidodendron vasculare* had, in the place of such tissue, large scalariform tubes, sometimes but not always mixed with orthosenchyma. Moreover, the central woody axis of *Sigillaria vascularis* was arranged in wedge-shaped masses, and penetrated by medullary rays, thus differing altogether from the entirely vascular zone constituting the woody axis of *Lepidodendron,* either *Harcourtii* or *vasculare.* The vascular bundles proceeding from the outside of the woody cylinder to the leaves, the zone of lax parenchyma gradually increasing in strength, and afterwards passing into elongated utricles or tubes, and having a radiating arrangement at its outside, as well as the epidermis, all appear to have been much the same in structure in both *Lepidodendron* and *Sigillaria.*

It appears from my specimens, hereinbefore described, that *Lepidodendron Harcourtii,* *Sigillaria vascularis,* and *Halonia regularis,* had all a similar mode of dichotomizing, and that the division of the pith and woody axis was much the same in these three plants. The facts are given as they appear from an examination of the specimens, and are left for the consideration of the physiological botanist, who is much abler than myself to investigate the system of dichotomization of ancient plants by the phenomena observed in living *Lycopodiaceæ.*

I have always had a doubt that *Lepidodendron* had the *Stigmaria ficoides* for its root, such as was proved to be the case with large, ribbed, and furrowed *Sigillariæ;* but I saw the probability of Mr. Dawes' view, that *Halonia regularis* might prove to be the root of *Lepidodendron,* both on account of its frequent bifurcations and on account of other characters, quite independent of the similarity in structure of the two plants.

The researches of Mr. Richard Brown and Professor Schimper led me to expect that *Lepidodendron,* as well as *Knorria,* had a Stigmarioid root. My own observations, and the specimens here described, lead me to conclude that *Halonia regularis* is the root of *Lepidodendron Harcourtii,* but not the root of *Sigillaria,* that being, as was before stated, *Stigmaria ficoides.* It is seen in the description of the specimens in this Monograph how

14

much *Stigmaria* resembles *Halonia* in its external appearance, with the exception of its tendency to bifurcate, which *Stigmaria* does not do after it has assumed its ordinary characters, and that it is very difficult, if not impossible, to distinguish an imperfect specimen of one of these roots from the other, although their internal structure differs as much as that of *Lepidodendron Harcourtii* does from *Sigillaria vascularis*. In short, it must be considered that *Halonia regularis* is merely a *Stigmaria* with its epidermis removed.

Now, when we compare the internal structure of *Halonia regularis* with that of *Lepidodendron Harcourtii*, we find they are the same in every particular. The medulla, composed of orthosenchyma, is the same in both plants; so also the interior and exterior of the woody cylinder, composed of larger and smaller barred vessels—the vascular bundles leading to the leaves—the lax and delicate parenchyma gradually increasing in strength until it passes into elongated utricles (prosenchyma), arranged in radiating series —and the epidermis forming the exterior, are all alike in *Halonia* and *Lepidodendron;* therefore, as to identity of structure, and so far as that is of any value, we are led to conclude that *Halonia regularis* is the root of *Lepidodendron Harcourtii*. Up to this time, however, I have not heard of their having been found absolutely united to each other, as was the case with *Stigmaria* and *Sigillaria*.

In my examination of the Lancashire Coal-measures I have generally found *Halonia regularis* associated with *Lepidodendron*. This is the case in the Peel Quarry, where both these fossil plants are found in about the same relative proportions. A similar result is obtained in the Upper Foot seam of Oldham and the Bullion Coal of Burnley and its vicinity; but in the Upper Brooksbottom seam *Halonia* occurs more frequently than *Lepidodendron*, so far as my experience in collecting the fossil plants in this seam of coal has extended. In many districts, no doubt, *Lepidodendron* is found in great abundance, with few or no traces of *Halonia regularis;* but in such cases we always meet with numerous ill-preserved specimens of *Stigmaria*. In my own experience well-defined specimens of *Halonia regularis*, showing the quincuncially arranged tubercles, are seldom found except in sandstones, and when the epidermis of the plant has been removed. The reason why they have not been recognized in shales and "binds" is, that they are there compressed with their epidermis on, and have been generally taken for remains of *Stigmaria*.

PLATE XIII.

Lepidodendron Harcourtii.

Fig. 1 (No. 31). A transverse section of a stem from the Coal-measures near Dudley. Magnified 3¾ diameters.

Fig 2. A longitudinal section (from the centre to the circumference) of the stem. Magnified 4 diameters.

In this and the following plates the same parts of the specimens figured are indicated by the same letters, as follow :—

a. The middle part, showing the central axis or pith, composed of orthosenchyma.

b. The large scalariform tubes forming the inner portion of the woody cylinder.

c. The smaller scalariform tubes forming the outer portion of the woody cylinder, from which the vascular bundles communicating with the leaves or rootlets proceed.

d. The vascular bundles, proceeding from the woody cylinder and extending to the leaves or rootlets.

e. The mass of parenchymatous tissue, at first (near to the woody axis) of a fine and lax character, but increasing in size and strength as it proceeds outwards, until it becomes prosenchymatous.

f. The elongated utricles or tubes, arranged in radiating series, forming the outer zone next the epidermis.

g. The epidermis, converted into coal.

Fig. 3. A longitudinal section of a portion of the stem, showing part of the pith, the inner and outer portions of the woody cylinder, the origin and structure of the vascular bundles, and the delicate parenchyma traversed by them. Magnified 7 diameters.

Fig. 4. A longitudinal section of a portion of the outer part of the stem, exhibiting the coarse parenchymatous tissue, gradually passing into prosenchyma, and thence into elongated utricles or tubes; and the epidermis. Magnified 12 diameters.

Fig. 5. A nearly right-angled transverse section of one of the vascular bundles, near the outside of the stem, displaying the vascular axis, surrounded by a zone of orthosenchymatous tissue, then a regular circle of fine parenchyma, and the mass of coarse parenchyma in which the whole is imbedded. Magnified 24 diameters.

Fig. 6. A transverse section of a part of the stem near the outside, showing the coarse parenchyma traversed by a vascular bundle, and a portion of the outer radiating cylinder, and the epidermis. Magnified 5 diameters.

Plate XIII

Fig 6

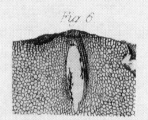

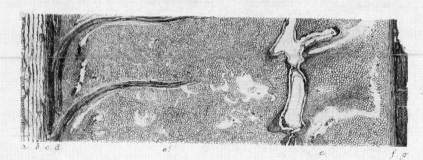

Fig 2

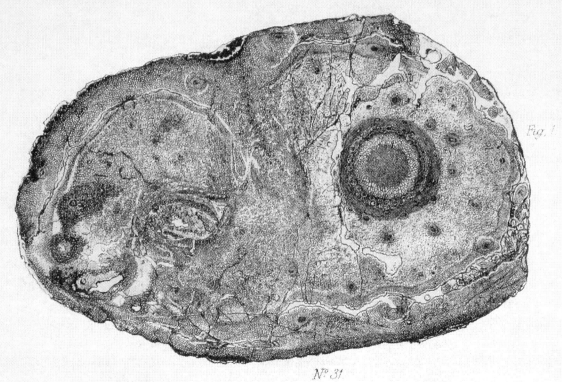

Fig. 1

N.º 31

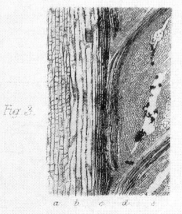

Fig 3.

Fig. 5

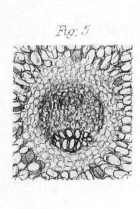

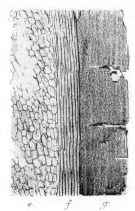

Fig 4

PLATE XIV.

Lepidodendron Harcourtii.

Fig. 1 (No. 32). A portion of a stem cut transversely, showing the rhomboidal scars and the leaf-attachments on the outside, and the two vascular axes of a horseshoe shape in the inside. This stem is cut across, probably at a point just before it commenced to divide in two. Natural size.

Fig. 2. One of the horseshoe-shaped bodies, showing vascular scalariform structure on its outside margin. Magnified $3\frac{1}{2}$ diameters.

Fig. 3. The other. Magnified $3\frac{1}{2}$ diameters.

Fig. 4 (No. 33). *Sigillaria vascularis.* A transverse section of the stem, showing two woody cylinders, with an apple-shape section, cut across at a point before the stem began to dichotomize. From the Bullion seam of Coal at Spa Clough, near Burnley. Magnified $3\frac{1}{2}$ diameters.

Fig. 5. Section of one of the central bodies, showing large detached (scalariform) vessels in fine orthosenchymatous tissue in the middle, and a woody cylinder of (scalariform) tubes or utricles, gradually diminishing in size as it goes outwards towards a radiating cylinder of barred tubes in wedge-shaped masses. Magnified 10 diameters.

Fig. 6. The other. Magnified 10 diameters.

Plate XIV

Fig 1.

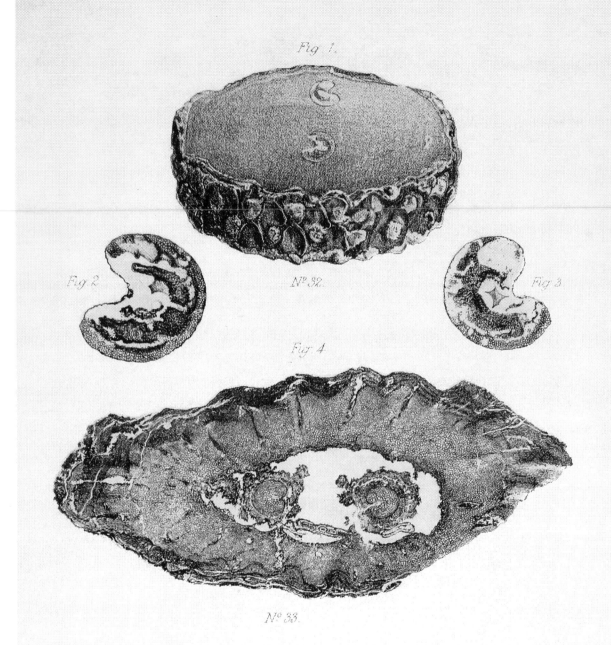

N° 32.

Fig 2.

Fig 3.

Fig. 4.

N° 33.

Fig 5

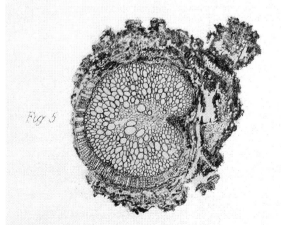

Fig 6

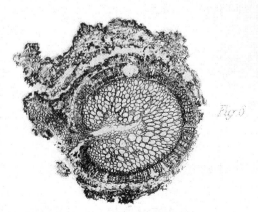

PLATE XV.

Halonia regularis.

Fig. 1 (No. 34) represents a transverse section of a root, exhibiting a disarranged woody cylinder and the mouths of numerous vascular bundles traversing the parenchymatous tissue, of which it was chiefly composed. From the Coal-measures near Dudley. Magnified $3\frac{1}{5}$ diameters.

Fig. 2. A longitudinal section of a portion, showing the medulla, composed of orthosenchyma, surrounded by a woody cylinder of scalariform tubes, from which proceed vascular bundles (several of them being displayed). Magnified 4 diameters.[1]

Fig. 3. A longitudinal section of a portion of the centre of the root, showing the medulla and the woody cylinder. Magnified 8 diameters.

Fig. 4. A transverse section (at nearly right angles) of one of the vascular bundles, near the outside of the root. Magnified 25 diameters.

---

[1] The artist has represented in fig. 2 the specimen as a stem, and not as a root, as we believe it to be. In order, therefore, to show the true direction of the vascular bundles leading to the rootlets, the figure should be reversed.

Plate XV.

Fig 2.

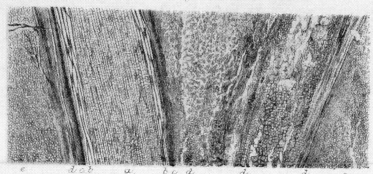

e   d c b   a   b c d   d   d   e

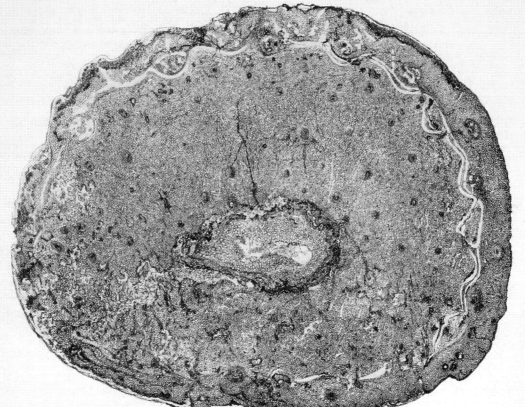

Fig 1.

N.° 84.

Fig 3.

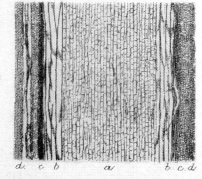

d c b   a   b c d

Fig 4.

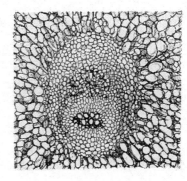

J.N Fitch del et Lith.   R Fitch imp

PLATE XVI.

Halonia regularis.

Fig. 1 (No. 35) represents a portion of a root, showing the tubercles, having a raised central pimple, and the woody axis in the inside of the root   From the Upper Brooksbottom seam of Coal, Lancashire.   Natural size.

Fig 2.   A transverse section of the root, showing the medulla, woody cylinder, a vascular bundle, portions of the parenchyma, and the outer radiating cylinder.   Magnified 5 diameters.

Fig. 3.   A longitudinal section of the root, representing the medulla, vascular cylinder, parenchymatous tissue, and outer radiating zone.   Magnified 10 diameters.

Fig. 4.   A transverse section of the medulla and woody cylinder.   Magnified 10 diameters.

Fig. 5.   A transverse section (nearly at right angles) of one of the vascular bundles in the inside of the root.   Magnified 40 diameters.

Plate XVI

Fig. 1

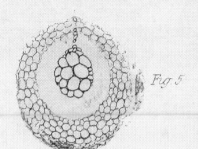

Fig 4

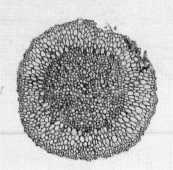

Fig 5

N.º 35

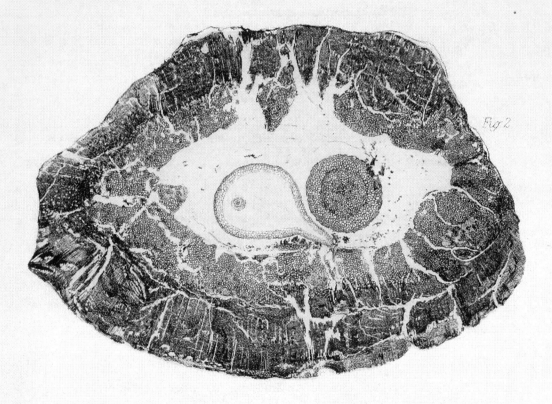

Fig 2

Fig 3

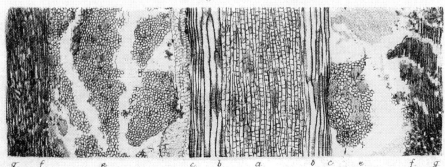

g f e   c b a b c e f g

J.N.Fitch del. et Lith.   R.Fitch imp.

PLATE XVII.

Halonia regularis.

Fig. 1 (No. 36)   A portion of a root, showing the depressed areolæ with a central raised pimple.   From the Upper Brooksbottom seam of Coal, Lancashire.   Natural size.

Fig. 2.   A transverse section of the same root, showing the medullary axis, the woody cylinder, the parenchyma (passing gradually into prosenchyma), and the epidermis. Magnified 5 diameters.

Fig. 3.   A transverse section of one of the raised papillæ in the areolæ, showing the vascular tubes surrounded by a space formerly occupied by orthosenchymatous tissue, and succeeded by a zone of fine parenchyma.   Magnified 20 diameters.

Fig. 4 (No. 37).   A forking portion of a root, from the Upper Brooksbottom seam of Coal, Lancashire.   Natural size.

Fig. 5 is a transverse section of the root, showing the medulla and vascular cylinder divided across, surrounded by the parenchymatous tissue, with the outer zone and the thick epidermis.   Magnified 5 diameters.

Fig. 6 is a transverse section of the medulla and the woody cylinder; the unshaded portion indicates their division.   Magnified 10 diameters.

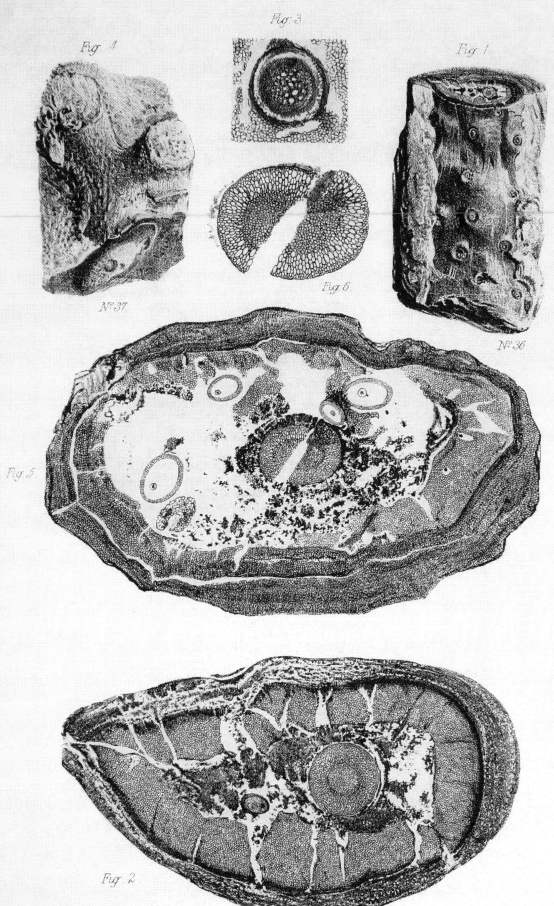

Plate XVII

Fig. 3

Fig. 4

Fig. 1

N.º 37

N.º 36

Fig. 6

Fig. 5

Fig. 2

J. N. Fitch, del. et Lith.

R. Fitch imp

PLATE XVIII.

Halonia regularis (No. 38).

This drawing represents the outside of a specimen, partly coated with its epidermis (converted into coal), and partly bared. Thus are seen both the areolæ and the tubercles, according to the condition in which the outside of the specimen is exposed. From the Peel Delph, near Bolton-le-Moors. Natural size.

Plate XVIII.

No. 38.

J.N.Fitch del et Lith.

R.Fitch imp.

THE

PALÆONTOGRAPHICAL SOCIETY.

INSTITUTED MDCCCXLVII.

VOLUME FOR 1875.

LONDON:

MDCCCLXXV.

OBSERVATIONS

ON THE

STRUCTURE OF FOSSIL PLANTS

FOUND IN THE

CARBONIFEROUS STRATA.

BY

E. W. BINNEY, F.R.S., F.G.S.

PART IV.

SIGILLARIA AND STIGMARIA.

Pages 97—147; Plates XIX—XXIV.

LONDON:
PRINTED FOR THE PALÆONTOGRAPHICAL SOCIETY.
1875.

PRINTED BY
J. E ADLARD, BARTHOLOMEW CLOSE.

CONTENTS.

		PAGE
I. Introductory Remarks		97
II. General Observations		98
III. Bibliography		104
§ 1. Steinhaur (Roots), 1818		105
2. Witham (Stem), 1833		108
3. Lindley and Hutton (Roots), 1838 . . .		108
4. Brongniart (*Sigillaria*, &c.), 1839 . .		108
5. Morris (Roots), 1840		112
6. Goeppert (Roots), 1841		113
7. Binney (Roots), 1844		113
8. Binney and Harkness (Roots), 1845 . .		113
9. Corda (*Diploxylon*), 1845 . . .		113
10. King (*Sigillaria* and *Anabathra*), 1845 . .		114
11. Binney (Roots), 1846		118
12. Brown (Roots), 1846		118
13. Binney (Roots), 1847		118
14. Hooker (Roots), 1848		118
15. Brown (Roots), 1848		118
16. Brown (Roots), 1849		118
17. Brongniart (*Sigillaria*, &c.), 1849 . .		118
18. Binney (Spores and Roots), 1849 . . .		119
19. Dawson (Roots), 1854 . . .		119
20. Goldenberg (Spores), 1855 . . .		119
21. Binney (Roots), 1858		119
22. Binney (Roots), 1861		119
23. Binney (*Sigillaria* and *Stigmaria*), 1862 . .		119
24. Binney (*Diploxylon*), 1865 . . .		121
25. Binney (*Sigillaria*), 1865 . . .		122
26. Binney (*Stigmaria*), 1865 . . .		127
27. Carruthers (*Sigillaria*, &c.), 1866 . .		128
28. Schimper (*Sigillaria*), 1870 . . .		128
29. Williamson (*Stigmaria*), 1871 . . .		131
30. Williamson (*Diploxylon*), 1872 . . .		131
31. Newberry (*Sigillaria*), 1873 . . .		134
32. Renault and Grand' Eury (*Sigillaria spinulosa*), 1874 . .		135
33. Brongniart (*Sigillaria*), 1875 . . .		136
IV. Description of the Specimens		136
§ 1. Nos. 39 and 40. *Sigillaria vascularis* . .		136
2. Nos. 41, 42, and 43. *Stigmaria ficoides* . .		139
3. No. 44. *Sigillaris vascularis* . . .		141
4. No. 45. *Stigmaria ficoides* . . .		143
5. Nos. 46 and 47. *Stigmaria ficoides* . .		144
V. Concluding Remarks		145

PART IV.

SIGILLARIA AND STIGMARIA.

I. INTRODUCTORY REMARKS.

In this Part of my Monograph it is intended to give a summary of the present state of our knowledge on the structure of *Sigillaria* and the allied plants, rather than descriptions of many new specimens. Further information, however, on *Sigillaria vascularis* will be given; and a *Stigmaria*, not only showing structure in the medulla, but in every respect agreeing with *Sigillaria vascularis*, will be figured and described. Some additional information on *Stigmaria* will also be furnished; and the structure of its rootlets, as well as the remarkable sutures dividing the base of the stem of *Sigillaria*, will be treated more at length than they have been in my previous papers.

Since the publication of my views, that large and small specimens of *Sigillaria vascularis*, as well as *Diploxylon cycadoideum*, had piths formed of barred tubes, and not parenchyma, several authors have doubted the correctness of the description given of my specimens, and asserted that such plants, as well as *Stigmaria*, had piths of parenchyma similar to those of the *Lepidodendron Harcourtii* described in this Monograph. More evidence will be adduced in support of my views, and it is hoped that it will be conclusively shown that at least one *Stigmaria*, and that the only specimen of the kind ever described with a pith in a perfect state of preservation, had a pith composed of barred tubes and cells. What was the structure of other *Stigmariæ*, so far as regards their piths, yet remains to be proved; and we must wait patiently for the discovery of specimens in such a state of preservation as to afford us the desired information. This may come any day, now such diligent search is being made for the discovery of Coal-measure Plants showing structure.

In the present as well as in the former Parts of this Monograph it has been my endeavour to describe the most perfect specimens that could be procured, showing the appearance of the exterior of the plant as well as its internal structure; as I have always found great difficulties in examining imperfect fragments of plants, however well their structure may have been preserved.

15

A great amount of labour, no doubt, has still to be devoted to the study of the structure of many of our common Coal-measure Plants before we can speak on all points with positive certainty.

II. General Observations on Sigillaria, Anabathra, Diploxylon, and Stigmaria.

Ever since the time when the Fossil Plants of the Coal-measures first attracted attention, *Sigillaria* has occupied a chief place in the minds of botanists; for it is to be met with in the strata near most seams of coal, in a more or less perfect state of preservation. The trunks of this genus are of two kinds, namely, those distinctly ribbed and furrowed, with leaf-scars on the ribs at greater or less distances, and those with the leaf-scars contiguous, and covering the whole surface of the trunk; both having them in a spiral arrangement around the axis. Nearly one hundred species have been described by different authors, who have made numerous species out of the same trunk; various parts of it being in a bad or good state of preservation. No doubt, when we are better acquainted with the true nature of the plant, the number of species will be greatly reduced.

For a long time *Sigillaria* and *Stigmaria* were regarded as distinct genera of plants, and even now, on the Continent, some distinguished palæontologists are disposed to remain of that opinion. In the specimens first described by me, in the 'Philosophical Magazine' for 1844,[1] which were found in Mr. Littler's quarry, near St. Helen's, *Stigmaria* was clearly traced to the trunks of the large, irregularly ribbed and furrowed *Sigillariæ*, showing little, if any, traces of leaf-scars; but it was there distinctly stated that around these trunks smaller trunks were found standing, which showed all the characters of *Sigillaria reniformis*, with *Stigmaria* rootlets in the adjoining strata, pointing in the direction of the root, but not absolutely proved to be connected with it. On viewing the specimens as they originally stood in the quarry before their removal, little doubt could be entertained as to all the trees there found having had *Stigmariæ* for their roots. In some specimens, however, afterwards described by me in the 'Philosophical Magazine' for 1847, ser. 3, vol. xxxi, p. 259, the connection of *Stigmaria*, as a root, with *Sigillaria reniformis, S. alternans,* and *S. organum,* was clearly proved.[2]

The regularly ribbed and furrowed *Sigillaria*, with distinct leaf-scars, generally found flattened and compressed in the sandstones and shales, are seldom of so large a size as

[1] 'Phil. Mag., ser. 3, vol. xxiv, p. 168; and 1845, vol. xxvii, p. 241, &c.
[2] See also 'Quart. Journ. of Geol. Soc.,' vol. ii, p. 391.

those irregularly ribbed and furrowed stems described by me under the name *Sigillaria vascularis*, sometimes attaining seven feet in diameter. In the fossil forests of trees standing erect in the Coal-measures, which have come under my observation, nearly all belong to the last-named genus. In the Pemberton Hill Cutting, on the railway between Wigan and Liverpool, six out of thirty stems, from one to two feet in diameter, exhibited the scars of *Sigillaria reniformis, S. alternans,* and *S. organum;* the remaining twenty-four belonging to *S. vascularis.* On the numerous fossil trees found in cutting the Clay-Cross tunnel, on the Midland Railway, near Chesterfield; in the specimens found in the deep pit at Pendleton, some of which were more than fifty feet in height; in that from the Victoria pit, Dukinfield, now in the Manchester Museum; in those on the Manchester and Bolton Railway, at Dixon Fold, described by Messrs. Hawkshaw and Bowman; and in the large stems from the Trap-Ash, of Laggan Bay, discovered by Mr. Wünsch; there was no evidence of distinct leaf-scars, but only irregular ribs and furrows. All the specimens except the last named were seen and examined by me *in situ.* The only example of a very large *Sigillaria* showing distinct leaf-scars, which has come under my observation, is specimen " No. 49 " of *Sigillaria reniformis,* now in the Museum of the School of Mines in Jermyn Street. Unfortunately, all the above-mentioned specimens, except those from Laggan Bay, afford no traces of internal structure. These last, however, some of which are about two feet in diameter, afford evidence of the structure of the thick inner bark, termed by me the outer radiating cylinder, and the woody or inner radiating cylinder of barred tubes, containing vascular bundles and medullary rays, enclosing a medulla, composed of barred tubes, in all respects exactly similar in structure to the large *Sigillaria vascularis,* with irregular ribs and furrows, described by me in the ' Philosophical Transactions;'[1] and the smaller specimens, exhibiting on their outsides scars of *Lepidodendron,* described in the ' Quarterly Journal of the Geological Society.'[2] These large and small specimens gradually pass one into the other, as numerous specimens in my cabinet, in addition to those figured, amply testify. Many persons have become accustomed to class my small specimens, the first ever described showing a medulla of vascular tubes, as *Lepidodendra,* from their external characters, without regarding their inner radiating cylinder and its singular medulla, so totally different in arrangement to the vascular cylinder and medulla of orthosenchymatous tissue of *Lepidodendron Harcourtii* before described in this Monograph.

When M. Brongniart described the structure of *Sigillaria elegans* he had before him, and described in the same Memoir as perfect specimens of *Lepidodendron* and *Stigmaria,* with the exception of the medulla and outer radiating cylinder, as have been met with up to this time; and he alludes to the probability of *Stigmaria* being the root of *Sigillaria;* but he notices the remarkable difference in structure between *Sigillaria* and *Lepidodendron.*

[1] For 1865, p. 579 *et seq.*　　　　　[2] Vol. xviii, 1862, p. 111.

The large *Sigillaria*, described in the 'Philosophical Transactions,' does not show many cicatrices of leaves on its outside, and is not of great size as a specimen of that genus, but it is the largest found by me in a calcareous nodule in a seam of coal. It is probable that it may have been portion of a main root, rather than a stem; for those portions of *Sigillaria*, whatever the characters of the stem, show nothing but irregular ribs and furrows on their surface. There are generally twenty-four of these main roots to one stem. In structure, however, it agreed with Brongniart's *Sigillaria elegans* more than any other then known plant; and it was classed with *Sigillaria* chiefly on that ground.

Owing to the small size of the nodules in coal, in which the fossil wood is found, we can never expect to find any very large specimens; for ten inches is the diameter of a very large nodule. Portions of *Sigillaria reniformis*, *S. alternans*, *S. catenulata*, *S. tesselata*, and *S. organum* have come under my observation, clearly showing the structure of the outer radiating cylinder or inner bark (first noticed in Brongniart's specimen), sometimes reaching to as much as five or six inches in thickness, and enveloped in a stout outer bark, converted into bright coal; but they are all destitute of the internal radiating cylinder and the medulla. The absence of the latter is what might have been anticipated, as it is so generally absent in *Stigmaria*; but why the former should not be met with is not so evident, except that in large trees, at the present day, decomposition commences in the centre, and extends towards the circumference; and so it may have been in ancient times. The tannin in the bark may have had greater power to resist decomposition than the inner parts of the tree.

In my figured specimens of small *Sigillaria vascularis*, the medulla is in a perfect condition; but in the large specimens of that plant, and in *Diploxylon*, described by me in the 'Philosophical Transactions' for 1865, the central portion of the medulla is somewhat disarranged. Since the publication, however, of that Memoir fresh transverse sections of the large specimen of *S. vascularis* have been made, which prove beyond question that the whole of the medulla is composed of barred tubes.

In my description of the inner radiating cylinders of the large specimens, mention is made of medullary rays of various breadths, some much narrower than the diameter of the tubes they traverse, and others considerably broader, corresponding with what Professor Williamson has since designated "primary and secondary rays." These are termed medullary rays or bundles[1] in the Memoir, and they chiefly relate to the primary rays, but there are also numerous medullary rays of one and two cells in breadth. They were met with in both *Diploxylon* and *Sigillaria vascularis*; and, although the divisions in the radiating cylinder of the former might appear to indicate that vascular bundles,

[1] In my paper I used the term "medullary rays or bundles," owing to the large rays being composed of vascular tubes and not of cellular tissue, as is generally the case in recent plants; but the smaller ones were of cellular tissue, like ordinary medullary rays. Objection may be made to the term, but in using it no hypothesis is advanced.

similar to what Corda had described in his specimen, had occupied them, none were seen. The rays, whether the large oval or the small ones, consisting of a series of single or double cells in a vertical line, were not distinctly shown in the longitudinal section, however large, but only in the tangential sections, which is rather singular. The larger or primary rays in the inner radiating cylinder were in no case absolutely traced to those traversing the outer cylinder, but in the small *S. vascularis*, figured in the woodcut No. 5 (p. 594), they were distinctly seen proceeding from the *outside* of the inner to the exterior of the outer radiating cylinder. Still, they were not absolutely proved to be connected with the inner primary rays. These latter, as previously stated, were only seen in tangential section, so it is difficult to speak with certainty whether they were composed of barred tubes or not.

Since the publication of my Memoir all my specimens of *Diploxylon* and *Sigillaria* have been again carefully examined in their longitudinal sections, and traces of vascular bundles like those so frequently found in common *Stigmaria*, and which form so marked a character in Corda's *Diploxylon cycadoïdeum*, have been found; but certainly not so distinctly, or communicating with the medulla, as shown in the transverse and longitudinal sections of his specimens.

In the outer radiating cylinder, or inner bark, the foliar bundles, enveloped in masses of very large and lax parenchyma, of a double-cone form, noticed by Professor King, are seen traversing the prosenchymatous tubes and pushing them aside; but these are shown chiefly in the tangential section, although a few traces of them are met with in the longitudinal section. One of the best examples hitherto met with is that figured in plate 34, fig. 2, of my Memoir in the 'Phil. Trans.' for 1865. These characters are much the same, whether observed in the large specimens, with irregular ribs and furrows, or those with rhomboidal scars on their outsides, like fig. 5, in plate 35, of that Memoir; thus showing, by their structure, that both specimens most probably belong to one plant. In the midst of this lax tissue the bundle of vascular tubes, in tangential section, presents a kidney shape, similar to what MM. Renault and Grand' Eury have noticed in *Sigillaria spinulosa*. In none of my sections, however, has there been seen any indication of the anastomozing observed by those authors in transverse and longitudinal sections of their specimen.

As the stems grow larger the lax cellular tissue enveloping the foliar bundles becomes less, so that in an outer radiating cylinder, five to six inches in breadth, little of it is seen; and what does appear is far more compact in structure than the very large cells of lax parenchyma seen nearer the centre. The wedge-shaped masses of parenchyma containing the foliar bundles of vascular tissue divide the wedge-shaped masses of prosenchymatous tissue; and these wedges have their thin and thick ends opposite to each other, the one increasing inwards and the other outwards. It is most probable, owing to the very large size of the cells of this lax parenchyma, that the space between the inner and outer radiating cylinders in *S. vascularis* is so often wanting in structure.

My neighbour Professor Williamson has described some portions of *Diploxylon,* *Sigillaria,* and *Stigmaria,*[1] which, he thinks, confirms his opinion that all these plants had piths composed of parenchyma, and not piths of vascular tubes of various sizes, and sometimes more or less mixed with orthosenchymatous tissue, as I had described as occurring in the two first-named genera; in fact, that their piths and the pith of *Lapidodendron Harcourtii* were much the same in structure.

My specimens described in the ' Quarterly Journal of the Geological Society' and in the ' Phil. Transactions' were probably in a more perfect state than any figured and described previously, so far as *Diploxylon* and *Sigillaria* were concerned. As for *Stigmaria* no one had described the pith except Goeppert[2] and myself; and both of the specimens described by us were looked upon as more than doubtful by recent writers. Professor Williamson says, "I have elsewhere called attention to the way in which the rootlets of *Stigmaria* have penetrated everything within their reach that was penetrable; and I have no doubt that in both Professor Goeppert's and Mr. Binney's specimens these supposed medullary vessels were really Stigmarian rootlets that had found their way into the interior of the cavity left by the decay of the medulla, and been mistaken for a part of the plant into which they had intruded themselves." Now, in my Staffordshire specimen, which exhibited all the external characters of *Stigmaria ficoides,* mention is made only of the large vascular bundles found in the axis, without calling them medullary or any other vessels. As figured in the plate and described in the letterpress no one could scarcely take them for the rootlets of *Stigmaria.* The woody cylinder was one of those having the inner parts of their circle close together, and not open, as in Professor Goeppert's specimen. It is possible that the large tubes in my specimen are not in their normal condition; and they may have been somewhat altered in the process of mineralisation; but it is very improbable that they were ever introduced into the axis after the pith had been removed. The specimen figured and described by Goeppert is very different from mine, being more open in the spaces between the wedges of the woody cylinder; and its central part is enclosed in a *Stigmaria,* showing the exterior in a most beautiful state of preservation. It appears to me that the vascular bundles in the pith, though it might be urged that they have been squeezed from their true position between the wedges of the inner radiating cylinder into the parts where they are now found, are certainly not intruded rootlets. In comparing Goeppert's and my specimens with Professor Williamson's any one will see that they are in a much more perfect state of preservation than the Oldham fragments.

Many beautiful specimens of *Stigmaria,* showing structure, have been met with in the trap-ash of Scotland by Mr. Greive and Mr. John Young of the Hunterian Museum, Glasgow, to both of whom I am much indebted for their kindness in presenting me with

[1] 'Phil. Transactions,' 1872; Part II of the Professor' Memoir "On the Organisation of the Fossil Plants of the Coal-measures," p. 215.

[2] 'Genres des Plantes fossiles,' pl. 13.

specimens. But all these have lost their piths, and fail to give us any information as to what they were composed of.

Some years since I found a specimen in the " Bullion " seam of coal at Clough Head, near Burnley, having every part of the medulla beautifully preserved; and, although showing little trace of the exterior of a *Stigmaria*, it affords undoubted evidence, in its transverse section, of several bell-shaped cavities, from which the rootlets proceeded. In every portion this plant resembles in structure the specimens of *Sigillaria vascularis*, formerly described by me in the ' Quarterly Journal of the Geological Society' and the 'Philosophical Transactions;' and, if perfect identity of structure is to be taken as proving the connection of root and stem, this must be held to be the root of that plant. It also shows the occurrence of barred tubes or utricles of very large size, almost as large as those found in my Staffordshire specimen, on which so much doubt had been thrown.

Another specimen of *Stigmaria* was found by me in the same locality as the last, having the open spaces between the wedge-shaped masses of wood freely communicating with the medulla, and not bounded by the dark line so marked in my Staffordshire specimen, and the one showing structure from Clough Head. Both specimens exhibit vascular bundles and medullary rays, traversing the woody cylinder alike; but the last-mentioned specimen has lost nearly all its medulla. It is now described for the purpose of showing its difference in structure from the first-mentioned specimen. In my paper on *Sigillaria* and *Diploxylon* in the 'Philosophical Transactions,' 1865, it is stated (p. 585) that " the lunette-shaped extremities of the inner radiating cylinder of *Diploxylon cycadoideum*, as well as those in my specimen, remind us of a similar arrangement shown to occur in *Stigmaria* by Dr. Hooker, Plate 2, fig. 14, 'Memoirs of the Geological Survey of Great Britain,' vol. ii, Part I; and they appear to differ from those found in *Sigillaria vascularis* in not being divided from the central axis by a distinct line of demarcation, just as the same author's *Stigmaria*, fig. 5, differs from fig. 14. The interior of the inner radiating cylinder of the former plant is more free and open, and not so sharp and compact as that of the latter plant. Indeed, from structure alone it would appear probable that the first-named *Stigmaria* was the root of *Diploxylon*, whilst the last one was the root of *Sigillaria vascularis*."

In the memoirs published on the structure of fossil plants it has always been stated by me that *Lepidodendron* was closely allied to *Sigillaria;* but, as previously mentioned in this Monograph, *L. Harcourtii* contains a medulla of orthosenchymous tissue, and no inner radiating cylinder; and, on the other hand, *Sigillaria vascularis* has a medulla composed of large and small vascular tubes, and an inner radiating woody cylinder. My opinion has been formed from an examination of my own specimens; and other authors may have reasons, unknown to me, for classing *Sigillaria vascularis* as a *Lepidodendron*.

M. Bronguiart, who has given to the world nearly all the knowledge we possess of the structure of *Sigillaria elegans*, never supposed that his plant was a *Lepidodendron*, however much he observed its resemblance in structure to *Stigmaria*, so as to induce him to believe that the latter plant was the root of the former. As previously stated, *Halonia regularis* is most probably the root of *Lepidodendron Harcourtii*, if identity of structure can prove it. The vascular bundles and medullary rays can often be distinctly seen in the transverse section of *Diploxylon cycadoideum*, and in the *Stigmaria*, with wide spaces between the woody wedges; but, so far as my knowledge extends, they have not yet been found in *Sigillaria vascularis* and the close-wedged *Stigmaria*, communicating with the outside of the woody cylinder next to the medullary sheath or the medulla itself. The last-named kind of *Stigmaria* has only been described by Dr. Hooker and myself; even so exhaustive an author as Professor Schimper does not figure nor allude to one. In my experience the one kind is nearly as common as the other.

Up to the present time little information has been published on the organs of fructification of *Sigillaria* with the exception of the cone described by me in the Philosophical Transactions,' 1865, p. 595. That specimen was very imperfectly figured in the woodcut (fig. 6); but from the form and arrangement of the bracts, and their resemblance to the form and arrangement of the leaf-scars of *Sigillaria organum*, I am strongly inclined to believe that it belongs to some species of *Sigillaria*.

The specimen described by Goldenberg does not appear to me to have belonged to a large-ribbed-and-furrowed *Sigillaria*. It is to be hoped, however, that some cones will soon be met with showing the structure of the central column to be the same as that of *S. elegans* or *S. vascularis*, as was proved to be the case with *Lepidostrobus* and *Lepidodendron* previously shown in this Monograph.

The specimen " No. 19 Cone," described at p. 49 of this Monograph, so far as structure goes, is the nearest to that of the stem of *Sigillaria vascularis* of any that have come under my observation; but that does not go so far as to prove perfect identity of structure. It affords little evidence of the characters of the spores; indicating microspores only.

III. BIBLIOGRAPHY.

In giving a summary of what has been published on *Sigillaria* it is only possible to quote the opinions of those authors who have written on the structure of the plant, without noticing the numerous writers who have described its external characters. The only exception to this rule is the insertion, at length, of the Rev. Mr. STEINHAUR'S Memoir on *Stigmaria*. This is so full and true a description of the root, its author being, too, the first to surmise that *Stigmaria* might be a root, that it appears desirable to give the paper at length. The views of the authors are generally given in their own words.

1. S<small>TEINHAUR</small>[1] (Roots).—" Sp. 1. *Phytolithus verrucosus*, plate iv, figs. 1—6. Martin, Petrificata Derbiensia,' plates 11, 12, and 13; Parkinson, ' Organic Remains,' vol. i, pl. 111, fig. 1.

" The fossil which has received this name from the ingenious author of the ' Petrificata Derbiensia ' is by far the most common, and perhaps the most remarkable, of this class. Woodward seems already to have collected numerous specimens, notwithstanding their bulk and comparative unsightliness (' Catalogue of English Fossils,' vol. i, part 2, p. 104; vol. ii, p. 59, &c.); and Mr. Parkinson has exercised considerable, though fruitless, ingenuity in elucidating them. It might appear presumptuous, after the labours of men of such distinguished abilities, to obtrude to public notice any further remarks, had not these authors left abundant room for observation, which place of abode and inclination have enabled the writer to pursue during a series of several years. Within this period we have collected several hundred specimens, worked many from the bed of clay in which they were embedded, and examined in quarries, on coalpit hills, among heaps of stone by the roadside, and in various other situations, several thousand. The geological situation of this fossil is well known to be the coal strata, in almost all which, as far as the writer is enabled to judge, it is found. Its geographical habitats in these strata may be partly collected from the works already quoted. The specimens more immediately examined were found in the neighbourhood of Fulneck, near Leeds, or in the space included by the towns of Leeds, Otley, Bradford, Halifax, Huddersfield, and Wakefield; but I have also found it on the top of Ingleborough; in the Coal strata of Northumberland; abundantly in Derbyshire; at Dudley in Shropshire; and in the neighbourhood of Bristol. With respect to mineralogical constituent matter, it seems always to coincide with that of the stratum in which it is imbedded, with a slight modification of density. It is most abundant in the fine-grained siliceous stone provincially called ' Calliard' and ' Ganister,' and in some of the coal ' Binds' or ' Crowstones,' which have probably received this appellation from spots of bitumen or coal attached to these petrifactions. It is rather less frequent in the beds of scaly clay or clay mixed with siliceous sand and mica; very common, but completely compressed, in the coal shales or bituminous slate-clay; of occasional occurrence in the argillaceous iron-stone; not rare in the common grit, and upper thick beds of argillaceo-micaceous sandstone, or rag, and sometimes, though rarely, discoverable in the coal itself. Mr. White Watson, of Bakewell, had also in his collection, which we examined, a specimen in the Derby Toadstone or Trap; and we have also noticed it in the limestone behind the Bristol Hot Wells, at its junction with the sandstone. So immense, however, is the number of relics that, when the eye has been accustomed to catch their appearance, it is scarcely possible to walk a furlong in the districts where they are at home without meeting them in one shape or another. The most perfect form in which this fossil occurs is that of a cylinder more or

---

[1] ' Transactions of the American Phil. Soc.,' vol. i, p. 265, 1868.

less compressed, and generally flatter on one side than the other (plate iv, figs. 1 and 2). Not unfrequently the flattened side turns in so as to form a groove. The surface is marked in quincuncial order with pustules, or rather areolæ, with a rising in the middle, in the centre of which rising a minute speck is often observable. From different modes and degrees of compression, and probably from different states of the original vegetable, these areolæ assume very different appearances ; sometimes running into indistinct rimæ like the bark of an aged willow ; sometimes, as in the shale impressions, exhibiting little more than a neat sketch of the concentric circles (figs. 4, 5, 6). Mr. Martin suspected that these pustules were the marks of the attachment of the penduncules of leaves, and pl. xii represents a specimen in which he thought that he had discovered the reliquiæ of the leaves themselves. We have examined the specimen when the drawing, which is extremely correct, was made, but are convinced that Mr. Martin was misled by an accidental compression in describing these leaves as being flat. Numerous specimens in ' Ganister,' in which the lateral compression of the trunk is generally trifling, place the assertion beyond a doubt that the fibrous processes, acini, spines, or whatever else they may be called, are cylindrical; and small fragments of these cylinders show distinctly a central line (pith) coinciding with the point in the centre of the pustule. Convinced of the existence of these fibres, we were soon able to detect their remains, forming consider-able masses of stone, particularly of Coal-bind, on Wibsey Slack and at Lower Wike, where their contorted figure imitates the figure of Serpulæ ; but it excited much surprise, on examining the projecting ends of some trunks which lay horizontally in a bed of clay, extending along the southern bank of the rivulet which separates the townships of Pudsey and Tong, and which is exposed by slips in several places, to find traces of these fibres proceeding from the central cylinder, in rays through the stratum, in every direc-tion, to the distance of above twenty feet. Repeated observations, and the concurrent conviction of unprejudiced persons made attentive to the phenomenon, compelled the belief that they originally belonged to the trunks in question, and consequently that the vegetable grew in its present horizontal position at a time that the stratum was in a state capable of supporting its vegetation, and shot out its fibres in every direction through the then yielding mud. For if it grew erect, even admitting the fibres to be as rigid as the firmest spines with which we are acquainted, it would be difficult to devise means gentle enough to bring it into a recumbent posture without deranging their position. This supposition gains strength from the circumstance that they are found lying in all directions across one another and not to any particular point of the compass.

" The flattened and sometimes grooved form of one side of the cylinder has already been noticed. Woodward already observed that along this side there generally, or at least frequently, ran an included cylinder, which at one extremity of the specimen would approach the outside, so as almost to leave the trunk, while on the other end it seemed nearly central. A reference to his ' Catalogue,' vol. i, part 2, p. 104, to Mr. Parkinson's ' Organic Remains,' vol. i, p. 427, and to Martin's ' Petrificata Derbiensia,' fig. 1 c, will show how much this included cylinder has embarrassed those who have considered it with a view to the vegetable organ to which it owes its origin. In the specimens of Calliard, which have suffered little compression, but which are seldom above a few inches in length, this body is generally nearly central; perhaps in no instance perfectly lateral. In the specimens in clay, from one of which we were able to detach upwards of 6 feet, the flattened or grooved side is invariably downward; and consequently the included cylinder is in the position which it would assume if it had subsided at one end, while the other was supported, or which would be the result of its sinking through a medium of nearly the same specific gravity with itself, provided it was at one end rather denser than at the other. It must be observed that this included body appears to have suffered various degrees of compression, being sometimes cylindrical, which was evidently its original form, and sometimes almost entirely flattened. In the coal shale we were never able to detect a trace of its existence.

" Besides these indications of organisation, we have met with several specimens which, on being longitudinally split, discovered marks of perforations or fibres, more or less parallel with the axis of the cylinder, and in some degree resembling the perforations of Terebellæ in the fossil wood of Highgate and some other places. Whether these configurations be owing to the organisation of the original vegetable, or to some process which it underwent during its decay, seems impossible to determine; the specimens examined afforded no opportunity of discovering a connection between these tubes and either the internal cylinders or the external surface.

" Among the vast number of specimens examined, only one was detected which appeared to terminate, closing from a thickness of 3 inches to an obtuse point. We have given a figure of it, pl. 4, fig. 3. Two instances also came to our knowledge of branched specimens, in which the trunk divided into nearly two equal branches. So rare an occurence of this circumstance would, however, rather induce the supposition that the original was properly simple, and that these were only exceptions or monstrosities. The size of different species varies greatly; but we have seen none under 2 inches in diameter; the general size is 3 or 4, and some occur, but with very indistinct traces of the pustules, even 12 inches across.

" From the above, it appears rational to suppose that the original was a cylindrical trunk or root, growing in a direction nearly horizontal in the soft mud at the bottom of fresh-water lakes or seas, without branches, but sending out fibres from all sides; that it was furnished in the centre with a pith, of a structure different from the surrounding wood or cellular substance, more dense and distinct at the older end of the plant and more similar to the external substance towards the termination, which continued to shoot; and perhaps that, besides this central pith, were longitudinal fibres proceeding through the plant, like those of the roots of the *Pteris aquilina*. With respect to any stem arising from it, if a root or foliage belonging to it, if a creeping trunk, we have hardly ground for a supposition."

2. WITHAM[1] (Stem) describes his *Anabathra pulcherrima*, found in the Mountain-limestone series at Allenbank, in Berwickshire, as follows :—" A medullary axis ; woody tissue consisting of elongated cellules; medullary rays scattered at great distances. Stems roundish or compressed, tapering; pith of irregular polygonal cellules ; woody tissues, in the transverse section, presenting the appearance of regular, parallel, radiating series of four-sided or subhexagonal cellules, with radiating tubular ducts interspersed at intervals.　In the longitudinal sections the cellules have all their walls very regularly marked with parallel straight lines or ridges.　The medullary rays, in their transverse section, are of an elliptical form, and composed of irregular reticulations."

3. LINDLEY and HUTTON[2] (Roots).—Describing Mr. Prestwich's specimen of *Stigmaria,* they state—" The transverse section exhibited a meshing, something like that of Coniferæ, but with no concentric circles, and with the medullary rays consisting rather of open spaces between the other tissue, than of common muriform tissue found in such places. The longitudinal section (fig. 2) presented an assemblage of spiral vessels, of a very tortuous and unequal figure, without any woody or cellular matter intermixed.

" These formed a cylinder, which was surrounded externally by a mass of inorganic mineral matter, upon which surface the peculiar markings of *Stigmaria* were preserved, and which enclosed a hollow cavity, altogether destitute of mineral deposit.

" It would therefore appear that *Stigmaria* was a plant with a very thick cellular coating or bark, surrounding a hollow cylinder, composed exclusively of spiral vessels and containing a rather thick pith ; and that the plates of cellular tissue, which preserved the communication between the bark and the pith, were of so delicate an organisation that they disappear under the mineralising process which fixed the organic characters of the wood."

4. BRONGNIART[3] (*Sigillaria,* &c.).—" En faisant abstraction des colorations diverses de la silice qui occupe les parties dans lesquelles le tissu est complétement détruit, on voit que cette tige est formée de deux cylindres de tissus plus résistants, et dont la texture est parfaitement conservée, cylindres qui ne sont pas concentriques l'un à l'autre ; l'un, tout-à-fait extérieur et superficiel, constitue une sorte d'écorce, et présente extérieurement les bases saillantes, ou mamelons rhomboïdaux, qui correspondent aux points d'insertion de chaque feuille ; le tissu qui le compose, et qui parait parfaitement continu, est cellulo-fibreux, très-fin et très-dense ; l'autre cylindre, intérieur, rapproché d'un côté du cylindre extérieur, en est séparé par un espace assez large sur un côté, étroit de l'autre, qui parait avoir été occupé par un tissu cellulaire délicat (pl. i, figs. 3, 4, *e e′*), dont il ne reste de trace que dans quelques points, et surtout près de la zone corticale ou extérieure ; ce tissu cellulaire est représenté pl. ii, figs. 1, 2, 3, *e′* ; *e′*, l'intérieur de ce même cylindre (pl. i, figs. 3, 4, *a a* ; pl. ii, 1 *a*) ne présente que de la silice amorphe, transparente ou

[1] 'On the Internal Structure of Fossil Vegetables,' p. 74, Edinburgh, 1833.
[2] 'Fossil Flora,' &c., by Lindley and Hutton, vol. iii, p. 47, 1838.
[3] 'Archives du Muséum d'Histoire Naturelle,' &c., tome i, p. 410, 1839.

opaque, incolore ou diversement colorée, mais qui a pris probablement la place d'un tissu cellulaire, analogue à celui dont il reste quelques traces entre l'écorce et le cylindre intérieur.

"Quant à ce cylindre creux, à cette sorte de tube excentrique, la disposition et la nature des parties qui le constituent méritent de fixer en premier notre attention, car il représente le système vasculaire ou ligneux de la plante.

"Il forme un cylindre parfaitement régulier, de 13 à 14 millimètres de diamètre intérieur, et d'un millimètre d'épaisseur, composé d'un nombre déterminé de faisceaux, tous parfaitement égaux et semblables, placés les uns à côté des autres, sans aucun intervalle appréciable dans la plupart des cas, mais distincts par la forme arrondie de chacun d'eux du côté intérieur, ce qui donne au bord interne, sur la coupe transversale, une forme festonnée.

"Il suffit d'un faible grossissement pour reconnaître que chacun de ces faisceaux est formé de deux zones distinctes, l'une interne, constituant ces sortes de festons, l'autre externe, beaucoup plus étendue. Ces deux zones, quoique immédiatement contiguës, se distinguent facilement par une modification dans leur aspect et dans leur coloration vers leur point de contact ; mais un plus fort grossissement rend bientôt compte des différences de leur organisation.

"Sur la coupe transversale (pl. i, fig. 4 *b*, et pl. iii, fig. 1 *b b'*), on voit que les parties internes des faisceaux, ayant la forme d'un segment de cercle dont la convexité est tournée intérieurement, sont formées entièrement par un tissu dont les parois ont la même épaisseur et le même aspect ; ce sont, comme on le verra plus tard, des vaisseaux[1] à parois rayées transversalement ou obliquement, ou même réticulées, dont les orifices, anguleux et irréguliers, sont disposés sans ordre, mais dont les plus grands (*b*) sont du côté du centre du cylindre, les plus petits (*b'*), au contraire, vers l'extérieur et appliqués contre la zone externe de ce cylindre vasculaire.

"Cette zone extérieure (pl. i, fig. 3, 4, *c* ; pl. iii, fig. 1, *c c'*) est formée par un tissu disposé avec une grande régularité, en séries rayonnantes, tantôt tout-à-fait contiguës, tantôt séparées par d'étroits intervalles, occupés par des rayons médullaires, dont le tissu est maintenant détruit. Les orifices des vaisseaux (car ce sont encore des tubes rayés qui constituent toute cette zone) dont chacune de ces séries est composée vont en diminuant vers l'intérieur, les plus petits (*c*) étant presque en contact avec les plus petits vaisseaux des faisceaux internes, et ces vaisseaux d'un petit calibre formant, par leur rapprochement, mais sans se confondre, la ligne de démarcation entre les faisceaux internes, composés de vaisseaux disposés sans ordre, et les faisceaux externes, dont les vaisseaux sont disposés en séries rayonnantes, séparées par des rayons médullaires.

---

[1] J'emploie habituellement le mot *vaisseaux* pour indiquer ce tissu, quoiqu'il n'ait pas, ainsi qu'on le verra plus tard, les caractères des vrais vaisseaux. Ces tubes sont plutôt des utricules très-alongés et communiquant entre eux par leurs ouvertures latérales, comme les tubes fendus des Fougères et les tubes poreux qui forment le bois des Conifères, que de vrais vaisseaux dont les cavités seraient continues.

" Ces faisceaux, par leur contact presque immédiat, et la manière dont ils se correspondent avec une régularité parfaite, sont dans les mêmes rapports que les faisceaux fibro-vasculaires qui constituent le bois, proprement dit, dans les plantes dicotylédones, et les faisceaux de trachées qui, dans ces mêmes plantes, sont placés à la partie interne de ces faisceaux ligneux, et constituent l'étui médullaire. Aussi, quoique ces faisceaux internes n'aient pas exactement l'organisation et la disposition des faisceaux de trachées de l'étui médullaire, leur position, relativement aux autres parties, étant la même, je les désignerai sous le nom de faisceaux médullaires, pour les distinguer des faisceaux plus extérieurs, qui ont la structure rayonnante de la zone ligneuse, ce que j'appellerai les faisceaux ligneux.

" En dehors de ces derniers, on voit encore de petits faisceaux, dont la coupe transversale est arrondie, qui sont complétement isolés des faisceaux ligneux, mais qui en sont tantôt très-rapprochés, tantôt un peu plus éloignés, et qui correspondent exactement au milieu de chacun d'eux, puis enfin quelques-uns plus éloignés et disposés avec moins de régularité. Ces faisceaux sont, comme les faisceaux médullaires et ligneux, composés d'un tissu uniforme, mais plus fin, irrégulier et sans disposition rayonnante ; ils me paraissent avoir été isolés dans le tissu cellulaire extérieur, et n'être probablement que des faisceaux détachés du cylindre vasculaire et se portant dans les feuilles, mais qui ne seront conservés que dans la partie voisine de l'axe ligneux, tandis que la partie qui traversait obliquement la zone celluleuse extérieure aura été détruite, soit avant, soit pendant la pétrification, en même temps que le tissu cellulaire qui les environnait.

" Si nous examinons, au moyen de coupes longitudinales (pl. iii, fig. 2), ces mêmes parties vasculaires, dont je viens d'indiquer les positions respectives, telles que nous les offre la coupe transversale de la tige, nous verrons que tous les tissus conservés, et dont nous avons vu les orifices dans cette coupe, sont d'une structure très-analogue et ne présentent que de légères différences, qui peuvent échapper au premier coup d'œil, mais qui ne sont pas cependant sans quelque importance.

" Les faisceaux internes ou médullaires (pl. iii, fig. 2 *b b'*) sont composés d'utricules tubuleux très-allongés, très-inégaux en grosseur, dont les plus petits, *b'*, sont extérieurs, et les plus grands, *b*, sont placés au côté interne ; ces utricules, disposés sans régularité, assez flexueux, sont non seulement différents par leur grosseur, mais aussi par leur longueur.

" Les plus petits sont en même temps beaucoup plus courts, et leurs deux extrémités, terminées en cônes obtus, se présentent assez souvent simultanément dans le champ du microscope.

" Les plus gros, au contraire, sont aussi beaucoup plus allongés, mais cependant on les voit aussi se terminer par une extrémité close et arrondie.

" Les parois de ces utricules ont un caractère commun, c'est que toutes sont marquées de stries transversales ou spirales, très-nombreuses et assez fines, mais très-variables, soit de l'un à l'autre, soit dans les diverses parties de l'étendue d'une même utricule.

" Les plus gros (pl. iv, fig. 1 *b*), et ceux dont les angles sont les plus prononcés, présentent en général des stries transversales, perpendiculaires à leur direction longitudinale ou peu obliqués, qui se réunissent entre elles dans les angles de ces utricules. Ils sont alors très-analogues aux vaisseaux rayés de beaucoup de Fougères et de Lycopodes, sauf quelques différences sur lesquelles je reviendrai plus tard.

"Dans d'autres utricules, généralement d'un moindre calibre, les stries ou raies sont beaucoup plus obliques, contournées en spirales, mais encore unies entre elles dans les points qui correspondent aux angles de ces utricules. Ces vaisseaux à raies obliques, $b'\, b'$, passent très-fréquemment à une disposition réticulée très-régulière dans la plupart des cas, qui semblerait produite par deux ordres de stries obliques en sens inverse, et se croisant de manière à former un réseau à mailles rhomboïdales, ou devenant hexagonales par l'inflexion régulière de ces stries. Avec un faible grossissement, et par conséquent des lentilles d'un foyer moins limité, ou peut croire d'abord que l'on voit simultanément les stries spirales appartenant aux deux faces opposées d'un même utricule ; mais un grossissement plus considérable prouve que ces stries obliques en sens inverse sont tracées sur une même paroi, à moins toutefois qu'elles ne résultent de l'application très-intime des parois de deux utricules différents juxta-posés. La manière dont les fibres transversales passent aux fibres obliques, celles-ci à des fibres réticulées irrégulièrement, puis régulièrement (pl. iv, fig. 4), me fait cependant douter que cette explication soit exacte, et me porte à croire que ces diverses modifications s'opèrent dans les parois d'un seul et même utricule.

"Les utricules les plus petits de ces faisceaux, ceux qui sont situés vers la partie externe, et qui sont aussi moins étendus en longueur, offrent encore une troisième modification (pl. iv, fig. 1 $b''\, b''$ et B), à laquelle cependant on arrive insensiblement. Ils présentent de véritables fibres spirales continues, au nombre de 2, 3, ou 4, se contournant parallèlement les unes aux autres, sans aucune réticulation, exactement comme dans les trachées à fibres multiples, sauf la plus grande brièveté des utricules qui présentent cette structure, et l'espacement sensible des tours de spires, qui peut faire penser qu'ils étaient unis par une membrane appréciable, et qu'ils se rapportaient par conséquent plutôt à la modification qu'on a designée sous le nom de fausses trachées.

"L'intervalle qui sépare ces fibres, soit dans ces utricules à fibre spirale, soit dans ceux à fibres obliques ou réticulées, soit enfin dans ceux à fibres transversales, ne varie pas sensiblement dans un même utricule, mais varie notablement de l'un à l'autre ; il est moindre dans les utricules d'un petit calibre, à fibres généralement en spirale, et atteint son maximum dans les plus gros utricules, à fibres transversales ou peu obliques ; mais ces variations sont comprises entre $\frac{1}{300}$ et $\frac{1}{400}$ de millimètre. Si ces utricules, allongés et striés en spirale, ne sont pas de vraies trachées, on voit cependant qu'elles ont beaucoup d'analogie avec ces vaisseaux par l'obliquité et la disposition spirale de leurs fibres, et sont, pour ainsi dire, intermédiaires entre les trachées à spire multiple et les vaisseaux striés des Fougères et des Lycopodiacées. . . . . .

"Ainsi, dans le *Lépidodendron Harcourtii*, dont j'ai pu examiner la structure dans tous ses détails, grâce aux échantillons qui ont été donnés au Muséum par M. Hutton et par M. R. Brown, on voit qu'il y a, comme dans le *Sigillaria elegans*, un cylindre vasculaire excentrique, séparé de l'écorce par une large zone d'épaisseur inégale, d'un tissu cellulaire en partie détruit, et renfermant une masse celluleuse centrale également très-altérée. Au premier aspect il semblerait donc y avoir beaucoup d'analogie entre ces deux tiges, mais un examen plus attentif montre que la structure du cylindre vasculaire est tout-à-fait différente.

"Dans le *Lépidodendron Harcourtii* il n'y a aucune trace de rayons médullaires, et le tissu vasculaire n'affecte pas cette disposition en séries rayonnantes, qui parait presque toujours être la conséquence de l'existence des rayons médullaires. Ainsi, par la disposition des éléments qui le constituent, le cylindre vasculaire de ce Lépidodendron n'a aucune analogie avec le cylindre ligneux des *Sigillaria*, des *Stigmaria*, ou des *Anabathra*, mais cependant il est formé d'éléments semblables, c'est-à-dire de ces tubes prismatiques rayés transversalement qui constituent le tissu ligneux ou vasculaire de ces trois tiges, et il semblerait représenter le cercle interne ou médullaire de l'*Anabathra* ou du *Sigillaria*, si dans ce dernier on supposait que les divers faisceaux qui le constituent fussent réunis en un cylindre continu.

"De même que nous avons remarqué que les tiges du *Stigmaria* avaient tous les caractères essentiels de celles du *Sigillaria*, si on supprimait dans cette dernière les faisceaux médullaires, de même on peut dire que le cylindre vasculaire continu du *Lépidodendron Harcourtii* représente la zone vasculaire intérieure ou médullaire de l'*Anabathra* (en admettant que nos prévisions sur la nature soient exactes), dépouillée de la couche ligneuse et épaisse qui l'environne. N'y aurait-il pas dans le premier cas simplement la différence d'une tige à une racine, dans le second d'un jeune rameau chargé de feuilles à une tige plus âgée? Cette dernière hypothèse me paraît cependant peu probable, à cause des prolongements vers l'extérieur que présente la zone vasculaire du *Lépidodendron Harcourtii* (pl. vi, fig. 5 *b'*; pl. vii, fig. 1 *b''*), prolongement dont on ne voit aucune trace sur la zone vasculaire interne de l'*Anabathra*."

5. Morris[1] (Roots).—In describing Mr. Prestwich's specimens Professor Morris says—"The portion of the stem of which the figure is a transverse section is in more perfect preservation than the specimens of *Stigmaria* usually are found in ; it is of nearly cylindrical form, about 4½ inches diameter, the external surface exhibiting the usual markings of this curious plant ; the internal part, with the exception of a vascular cylinder (also mineralised), being replaced by clay-ironstone."

Referring to the description of *Stigmaria* in the 'Fossil Flora,' Professor Morris states—"It has been thought advisable to have another section represented, with a view of showing what has hitherto not been well illustrated in the published figures of its structure. The internal cylinder in the specimen (fig. 3) is concentric, and consists of

[1] "On the Geology of Coalbrook Dale," by Joseph Prestwich, 'Transactions of the Geological Society of London,' 2nd series, vol. i, Explanation of Plates.

wedge-like portions of vascular tissue, the rounded origin of which, internally, is well defined; these wedges are generally of equal or nearly equal size, but they occasionally become confluent by the joining of two or more of them together. The form of the space necessarily left, or interstices between the sections where these are distinct, varies a little, in some cases being of nearly equal breadth throughout, and in others becoming narrower outwards and appearing to terminate or contract about the middle of the vascular tissue, beyond which they again frequently widen outwards: these spaces often contain portions of oblique and smaller vascular cords, apparently arising at different depths in the vertical cylinder, the origin and connection of which with the cylinder is shown in the oblique section, where a single series of vessels is seen passing from it surrounded by tissue of smaller diameter (pl. xxxviii, fig. 3 *a*).

"In no specimen yet examined has the course of the oblique cords been absolutely ascertained, but there can scarcely be any doubt, as suggested by Mr. Brown (to whom we are also indebted for the above observations), that these vessels, after arising from the cylinder, passed to the tubercles of the surface, through the thick cellular tissue which once probably occupied the larger space in the original plant. The discovery of these smaller oblique vessels is an interesting feature in the anatomy of *Stigmaria*; and they have also been pointed out by Mr. Brown as existing in *Anabathra*; and one of these is actually figured by Mr. Witham in his work (pl. viii, fig. 10), but considered by him (p. 41) as a section of a medullary ray. The analogous vessels existing in *Lepidodendron Harcourtii*, as figured by Mr. Witham ('Trans. Nat. Hist. Soc. Newcastle,' 1832), appear to arise from the outer part of the vascular cylinder. A somewhat similar division is found in that division of *Lycopodiaceæ* consisting of *Psilotum* and *Tmesepteris*; in those genera (according to Brongniart, 'Veg. Foss.,' vol. viii, pp. 44, 45) the vascular cylinder from which the oblique cords proceed includes a central pith."

6. GOEPPERT[1] (Roots) describes a very good specimen of *Stigmaria* with the pith containing vascular bundles, interspersed in cellular tissue, and the structure of the rootlets having an axis of vascular tubes, surrounded by cellular tissue. He appears to have been the first author to publish information on the structure of those portions of the root.

7. BINNEY[2] (Roots) describes the Fossil Trees having *Stigmaria* roots in Littler's quarry, near St. Helen's, Lancashire.

8. BINNEY and HARKNESS[3] (Roots).—Further observations on the last-named specimens.

9. CORDA (Diploxylon).—Corda[4] describes the characters of his Diploxylon cycadeoidum:—"Truncus medullosus cylindricus; decorticatus extus longitudinaliter obscurè-

[1] 'Les Genres des Plantes fossiles,' Bonn, 1841.
[2] 'The London, Edinburgh, and Dublin Phil. Mag.,' ser. 3, vol. xxiv, p. 165, 1844.
[3] *Ibid.*, vol. xxvii, p. 241, 1845.
[4] 'Beiträge zur Flora der Vorwelt,' Prague, 1845.

striatus. Corpus corticale crassum medullosum. Cylindricus lignosus minutus, e stratis duplicibus compositus. Stratum internum continuum annuliforme, externo adpressum, vasis irregulariter positis amplis, sexangularibus. Stratum externum crassum, e vasis minutis seriatis et fasciculatim junctis compositum, et radiis vasorum ligni interni percursum. Radii medullaris nulli. Medulla ampla."

By the author's figure the medulla appears to be altogether wanting in the centre of the specimen; what it was composed of there is no evidence to show; it might have been parenchymatous tissue, or barred tubes. The vascular bundles shown traversing the woody cylinder, and their free communication with the inside medulla, and the rounded ends of such cylinder as well as the curved bundle of vascular tubes seen in the longitudinal section, show a considerable difference in structure from both *Diploxylon* and *Sigillaria vascularis* described by me in the 'Philosophical Transactions.'

10. KING (*Sigillaria* and *Anabathra*).—Professor W. King,[1] in a most excellent paper entitled "Contributions towards establishing the General Characters of the Fossil Plants of the Genus *Sigillaria*," gave a lucid account of Brongniart's *Sigillaria*, showing its connection with *Stigmaria*; also some valuable original observations on specimens from Ouseburn, and North Biddick, proving that *Sigillaria* had a Stigmaroid root. His remarks on Mr. Witham's *Anabathra pulcherrima* show that its woody cylinder was identical in structure with that of *Sigillaria vascularis*, having been furnished with the sharp dark line separating it from the pith, without the interstices between the wedge-shaped bundles of the woody cylinder so distinctly shown in the Corda's *Diploxylon cycadoïdeum* :—

" Let us, in the next place, consider that remarkable fossil which Mr. WITHAM was the first to make known, under the name *Anabathra pulcherrima*. At the time when *Anabathra* was described few botanists had attended to the minute difference in vegetable tissue, which forms so conspicuous a feature in the phytological works of the present day; hence a few errors have been committed in drawing up the description which has been published of this fossil. Some of these errors have been rectified by M. Brongniart in his 'Observations on the Internal Structure of *Sigillaria elegans*;' but, as there are others which this gentleman had not the means of correcting, I have been induced to enter into the following description more minutely than would have been otherwise necessary. It requires also to be stated that, with the view of enabling me to become acquainted with the internal structure of fossil plants in general, Mr. Witham has, in the most handsome manner, placed in my hands the whole of his invaluable collection of sections, among which there is an instructive suite of *Anabathra*. To this gentleman, for so marked an act of kindness, there is certainly due from me an expression of deep obligation.

" Before commencing to describe the tissues of *Anabathra* it is necessary to make a slight reference to the state in which Mr. Witham's specimen existed when first dis-

[1] 'Edinburgh New Philosophical Journal,' vol. xxxviii, p. 119, 1845.

covered. It was invested with an irregular coat of mineral matter; in which were observed numerous small portions of vegetable tissue, intermixed with what appear to be twigs. Mr. Witham has represented this coat, charged with its vegetable fragments, in pl. viii, fig. 7, of his ' Internal Structure of Fossil Plants.' The matrix, as it ought rather to be called, was in immediate contact with the tissue of what we shall presently see is the ligneous zone of the fossil—a circumstance which prevents us coming to any conclusion as to the thickness of its bark ; for instance, whether it was thin, like most of the Conifers, or thick, as is the case with the *Sigillariæ*, the Cycases and Cactuses. Mr. Witham, in his description, says that the specimen when complete was a tapering body several inches in length, rounded at the extremity, and resembling the termination of a stem or branch. In another part it is stated that the specimen, divested of its envelope, was compressed so as to have one diameter about a half greater than the other. ' At the lower part the large diameter was upwards of two inches ; and at the extremity one diameter is about half an inch, the other nearly a fourth.' I may observe that the sections before me answer to these and the intermediate sizes. If we were uncertain that *Anabathra* possessed a thick bark, there is something in the description just quoted which would induce one to suppose that this fossil was a short fleshy plant, resembling some of the Cactuses. Let it be understood, however, that I am far from thinking that this was the case. Mr. Witham states that the specimen presented the appearance of natural joints at the distance of about two inches, and that its surface was slightly striated in the longitudinal direction. I mention these circumstances merely to give it as my opinion that the striated appearance was caused by the very elongated tubes of the ligneous zone, and that the joints were simply transverse cracks.

" A very singular result has been brought about by mineralisation in Mr. Witham's specimen. A large portion of the radiated tissue has been destroyed ; what remains is contained in a narrow marginal strip and in numerous isolated pea-shaped bodies, imbedded in a crystalline matrix, and situated inwardly to the latter. The reader is therefore requested to fill up in imagination all the vacant spaces which are represented in figs. 2 and 3 of pl. iv, and with the same kind of tissue as that which forms the marginal strip and the isolated bodies. To aid this a transverse restoration of the vascular and ligneous system is given in fig. 1, which is a little above the natural size.

" *Anabathra pulcherrima* is undoubtedly a dicotyledonous plant. It possesses a broad ligneous zone (*a*, fig. 1, pl. iv), a large medullary sheath in the shape of a hollow cylinder (*b*), and apparently a large pith (*c*). The ligneous tissue consists of very much elongated tubes which are occasionally quadrilateral, but generally hexagonal ; they are arranged in radiating series, and are remarkably regular in diameter throughout the thickness of the zone, till within the precincts of the vascular cylinder, where they become considerably reduced. The apertures caused by sectionising these tubes are distinctly seen with a common magnifier. Their length appears to be considerable, since a longitudinal section nearly half an inch long shows none of the tubes with both terminations (see figs. 3 and 4). The whole of their walls are marked with fine trans-verse lines or bars, which in general are parallel to each other; but occasionally they divide as is represented in fig. 5. All the tubes have their walls of a uniform thickness, so that *Anabathra* displays no appearance of the concentric rings which are found in the wood of ordinary exogenous trees. The ligneous zone appears to have been intersected by numerous narrow medullary rays, judging from the interspaces which are marked *d* in figs 2 and 4.

"The vascular cylinder is composed of elongated tubes which on the transverse section are irregularly angular, and somewhat variable in their proportion. Those of the greatest diameter are a little larger than the tubes composing the marginal strip of the ligneous zone, and they constitute the inner four fifths of the cylinder; while the smallest, into which the others gradually pass, occupy the remaining or outer portion. At the margin of the cylinder the vessels have become so diminished in size as to resemble the small ligneous tubes which immediately circumscribe them; occasionally a small vessel is to be seen among the larger ones. With the exception of their being placed somewhat according to size, as just stated, the tubes of the medullary sheath possess no order in their arrangement. The tissue of this part appears to be shorter than that of the ligneous zone, as there are several terminations displayed on a longitudinal section (see *b*, fig. 3); but I am strongly inclined to believe that the shortness is more apparent than real: it ought rather to be said that the tubes in their longitudinal direction are very flexuous and twisted round each other. This circumstance, by causing a longi-tudinal section to display certain of the tubes obliquely cut, and others deviating from each side of the plane of the section would produce, it is conceived, the appearance as if these cuts and deviations were so many terminations. The walls of the tubes are marked with transverse lines or bars, which differ somewhat from those on the ligneous tissue, inasmuch as they are closer to each other, and they are often seen coming in contact, which gives them an anastomosed appearance (see fig. 6, pl. iv). In none of the large vascular tubes are the lines so disposed as to form a spiral, either broken or continuous; probably this is the case in the smallest, but the section is not sufficiently thin to allow of its being seen. The vascular cylinder is in close contact with the ligneous zone; and in no part does it display the least appearance of openings or medullary rays.

"The pith appears to have been formed of fusiform cells, analogous to those which Brongniart describes as belonging to the corresponding part of *Lepidodendron*. It may be doubted, however, that what I have considered as forming a portion of the pith of *Anabathra* did in reality belong to this part, since it is simply a portion of fusiform tissue crossing the centre of one of the transverse sections.

" Reverting to the ligneous tissue, and adverting to the longitudinal section repre-sented in fig. 4, pl. iv, which is at right angles to the medullary rays, and through the marginal strip, our attention must now be directed to these large openings (*e*) which form so prominent a feature. There are only two represented, owing to a greater number requiring more space than could be allotted for the figure; it consequently requires to be stated that they are arranged in a spiral manner. Mr. Witham described these openings as containing the medullary rays, which is not the case, because what has been probably taken for cellular tissue is in reality a number of small vessels (*f*) similar to those which occupy the outer part of the medullary sheath. Although the longitudinal sections do not exhibit any of these bundles springing from the vascular cylinder, their proximity to this part in some transverse sections (see fig. 2), together with the fact just stated, leaves no room to doubt their having constituted the leaf-cords of the plant. According to Mr. Morris it would appear that Dr. Brown had ascertained this point some time since.[1] Owing to one of the openings, or vascular passages, having been intersected in a portion of its course through the ligneous zone, as shown in the longitudinal section parallel to the medullary rays in fig. 3, pl. iv, we have displayed in a very instructive manner a leaf-cord, or vascular bundle (*f*), traversing at right angles the ligneous tissue; a similar bundle is exhibited in the transverse section (fig. 2). These two sections prove that the leaf-cords curve but very slightly in their passage through the ligneous zone, as they proceed horizontally for a considerable distance. From the passage being in part hollow (see fig. 4), it may reasonably be supposed that the cords were accompanied in their course with a portion of cellular tissue.

"We may now be permitted to say a few words on the comparative anatomy of *Anabathra*. No one can help being struck with the similarity which this plant possesses in some points of its structure to *Sigillaria* and *Lepidodendron*. The width of the ligneous zone is certainly greater in *Anabathra* than in *Sigillaria*; but there scarcely appears to be a shade of difference in the character of its constituent tissue in either plant; while between *Lepidodendron* and *Anabathra* there is in their vascular cylinder the closest resemblance. It is, therefore, clear that these three plants are nearly related to each other.

"The resemblance between *Anabathra* and *Lepidodendron* in their vascular cylinder has induced Brongniart to hazard a question to this effect: May not the latter be the young branch, and the former the stem, of one and the same plant? 'The hypothesis involved in this question,' says its author, 'appears, however, to have little probability in its favour, in consequence of there being on the outer part of the vascular cylinder of *Anabathra* none of the prolongations which are visible on the corresponding part of *Lepidodendron*.' The prolongations here alluded to are those portions of the leaf-cords which are on the point of curving off from the cylinder, to the margin of which they give a sinuous appearance. Mr. Witham's transverse sections of *Anabathra* certainly do not show any sinuosities. Brongniart's objection is, therefore, so far a valid one; but it seems to me that, before *Lepidodendron* can be considered as the branch of *Anabathra*,

---

[1] 'Transact. Geological Society,' 2nd series, vol. v, description of pl. xxxviii. See above, p. 113.

there is required to be known an example of a Dicotyledonous tree having young
branches without any radically arranged ligneous tissue.

"*Sigillaria elegans* possesses in its anatomy a peculiarity of considerable interest in
a physiological point of view. It is furnished with a medullary sheath, which, there is a
strong reason to believe, existed to a certain degree independently of the ligneous
zone. But whatever doubt might stand in the way of such a peculiarity possessing
itself of our own conviction, so far as *Sigillaria* is concerned, it is clearly demonstrated
by what is observable in *Lepidodendron* and *Anabathra,* inasmuch as in the former the
vascular cylinder has performed its function without the presence of the ligneous zone of
the latter; add to this, that in *Anabathra,* although these two parts are in immediate
contact with each other, the differences which have been pointed out in their respective
tissues further prove that they represent independent systems. It will now be seen on
what grounds the distinction has been made in this paper between the vascular and the
ligneous part of the fossils which have been mentioned."

11. BINNEY[1] (Roots).—Description of the Dukinfield *Sigillaria* with long *Stigmaria*
roots, showing how the latter change their characters as they extend outwards from the tree.

12. BROWN[2] (Roots.)—In a paper on a group of fossil trees, in the Sydney Coal-
field of Cape Breton, with *Stigmaria* roots.

13. BINNEY[3] (Roots).—Description of *Sigillaria reniformis* having *Stigmaria ficoides*
as its roots, from the Pemberton Hill Cutting, on the Bolton and Liverpool Railway.

14. HOOKER[4] (Roots).—Dr. Hooker describes *Stigmaria* with close wedges in the
woody cylinder, as well as one with open wedges, and shows that these two different roots
belong to specimens having the same external characters. He also notices the depth of
the bell-shaped cavities from which the rootlets proceed.

15. BROWN[5] (Roots).—Description of an upright *Lepidodendron* with *Stigmaria*
roots in the Sydney Main Coal, in the Island of Cape Breton.

16. BROWN[6] (Roots).—Description of a *Sigillaria,* with conical tap-roots found in
the Sydney Main Coal.

17. BRONGNIART[7] classes under *Dicotylédones gymnospermes* the family *Sigillariées,*
containing the genera *Sigillaria, Stigmaria, Syringodendron, Diploxylon, Ancistrophyllum?* and *Didimophyllum?* and considers that Witham's *Anabathra* and Corda's
Diploxylon belong to the same genus. The same author, in treating on the *Diploxylon*
of Corda, writes:[8] "Ce genre n'est connu que par sa structure interne, qui me parait se
rapprocher du *Sigillaria,* dont il diffère cependant par le cylindre continu, formé par

[1] 'Quarterly Journal Geol. Soc.,' vol. ii, p. 391, 1846.
[2] *Ibid.,* p. 393, 1846.
[3] 'Phil. Mag.,' s. 3, vol. xxvii, p. 259, 1847.
[4] 'Memoirs of the Geological Survey,' vol. ii, part ii, p. 431, &c., 1848.
[5] 'Quarterly Journal Geol. Soc.,' vol. iv, p. 46, 1848.   [6] *Ibid.,* vol. v, p. 354, 1849.
[7] 'Tableau de Végétaux fossiles,' p. 97, 1849. (Ext. Dic. univ. d'Hist. Nat.)
[8] *Ibid.,* p. 57.

les vaisseaux qui environnent la moëlle et, suivant M. Corda, par l'absence de rayons médullaires.   M. Corda ne rapporte à ce genre qu'une seule espèce, le *Diploxylon cycadoïdeum*, décrite par lui, et trouvée dans le terrain houiller de Chomle, en Bohème ; mais je crois que c'est à ce même genre qui appartient, sans aucun doute, l'*Anabathra pulcherrima* de Witham (' Int. Struct. of Foss. Veg.,' p. 40, pl. 8), et je me fonde pour cela sur d'excellentes coupes de ce fossile remarquable, qui m'ont été adressées par ce savant, et qui montrent que le tissu qui entourne la moëlle détruite, mais dont on voit quelque trace, forme un cylindre continu sans direction rayonnante, et composé de vaisseaux rayés disposés comme dans *Diploxylon*.   C'est une seconde espèce de ce genre, à moins qu'on ne croie·devoir réserver à ce groupe le nom d'*Anabathra*."

18. BINNEY[1] alludes to the occurrence of spores in the inside of *Stigmaria*, and notices the remarkable crucial sutures on the base of the stems of some *Sigillariæ*.

19. DAWSON[2] (Roots), in his description of the Coal-measures of South Joggins, Nova Scotia, alludes to *Sigillaria* having *Stigmaria* roots.

20. GOLDENBERG[3] describes and figures spherical bodies, some with a triradiate ridge, and others without that character, as the fruit of *Sigillaria*, *Stigmaria*, and fossil *Selaginæ*.   These bodies, according to the author, appear to be attached to the scales of the cones, and not contained in a sporangium; and in the figures they appear chiefly at the base of the specimen.

21. BINNEY[4] (Roots) gives information as to the origin of the medullary rays, and the nature of the vascular bundles, in the pith of *Stigmaria*, also as to the structure of its radicles.

22. BINNEY[5] on *Sigillaria* and its roots.

23. BINNEY[6] (*Sigillaria* and *Stigmaria*):—" In the present paper it is my intention to confine myself to the description of three specimens of fossil plants which would generally have been designated *Lepidodendron* in England and *Sagenaria* on the Continent.

" No. 1.—The specimen illustrated in pl. iv consists of a cylindrical stem $\frac{8}{10}$ths of an inch in diameter, nearly enveloped in its stony matrix, and only showing its external characters on one side.   These consist of rhomboidal scars of an elongated and somewhat irregular form, arranged in quincuncial order, but not so perfectly as seen in most species of *Lepidodendron*.   In the middle of each scar there is an oval depression, from which rises a rounded prominence, where the leaf was attached.   The scars resemble those of *Lepidodendron selaginoides*, figured by Messrs. Lindley and Hutton in their ' Fossil Flora,' vol. i, fig, 12, but the depression in the scar on their specimen is not so marked as in mine.

[1] ' Quart. Journ. Geol. Soc.,' vol. x, p. 1, 1854.
[2] Ibid., vol. vi, p. 17, 1849.
[3] ' Flora Saræpontana fossilis,' pl. B, figs. 18 to 25 (1855), pl. x, figs. 1 and 2 (1857).
[4] ' Quart. Journ. Geol. Soc.,' vol. xv, p. 76, 1858.
[5] ' Trans. Manchester Geological Soc.,' vol. iii, p. 110, 1861.
[6] ' Quart. Journal Geol. Soc.,' vol. xviii, p. 107, &c., 1862.

" In the middle of the large cylinder last described is a smaller one, about $\frac{1}{7}$th of an inch in diameter. This is composed of large hexagonal vessels, of irregular sizes (*a, a*), placed one beside the other, without order, but becoming smaller as they approach the circumference, all having their sides barred with transverse striæ, and some of the smaller ones (*a, a*) being divided at short intervals by horizontal and oblique partitions. The outside of this inner cylinder[1] (*b, b*) is composed of hexagonal cells, barred with transverse striæ, about $\frac{1}{6}$th of the diameter of those contained in the centre, arranged in radiating series of a wedge shape, and divided by medullary rays or vessels very finely barred (*c, c*) as in the vascular cylinders of *Sigillaria* and *Stigmaria*, respectively described by Brongniart and Hooker. Around and placed next to the cylinder are a number of round bundles of fine vascular tissue (*d, d*), some of which are opposite to the medullary rays or vessels, and others apparently away from them, near the wedges of the wood. These bundles seem to be connected with the vessels which supply the leaves, but cannot be well traced to the medullary rays in all cases. It is probable that they may be sections of vessels passing from the medullary rays, or vessels, to the leaves. They are evidently the same vessels as are figured by Messrs. Lindley and Hutton ('Fossil Flora,' vol. ii, Pl. 99, fig. 1), and also resemble the vessels described by Brongniart as occurring on the outside of the woody cylinder in *Sigillaria elegans*. On the external portion of the outer radiating cylinder of the specimen similar vessels can be distinctly traced into the projecting scars from whence the leaves arise.

" Next occurs a space of about $\frac{4}{10}$ths of an inch (*e, e*), in which the tissue has for the most part disappeared and been replaced by mineral matter; but it seems to have been composed of delicate cellular tissue, which was traversed by bundles of vessels leading from the axis to the leaves. Then comes a zone of coarse cellular tissue (*f, f*) which gradually passes into small elongated utricles, of hexagonal form, and arranged in radiating series, which probably formed the inner bark. These in their turn pass into a black carbonaceous matter (*h, h*), the remains of the outer bark of the tree. The vessels traversing the external cylinder are of the same character as those traversing the internal one, except that they are of much greater size, each of the latter being probably composed of two or more of the former, as Dr. Hooker describes in *Sigillaria*.[2] A transverse section of the specimen ' No. 1 ' is similar to the same section of *Sigillaria elegans*, with this exception, namely, that the inner lunette-shaped bundles of vessels found within and next to the woody cylinder in M. Brongniart's specimen fill the whole of the central axis in mine. At first sight it might have been supposed that the specimen of *Sigillaria elegans* beforenamed had some of its middle portion destroyed, and that the lunette-shaped bundles once occupied the whole of the central axis ; but having by the kindness of M. Brongniart been permitted to examine the original specimen preserved in the

---

[1] " In this specimen by some cause a portion of the inner cylinder has been destroyed, either by the section not being cut true or by a part of the woody cylinder having been destroyed in calcification."

[2] " ' Memoirs of the Geological Survey of Great Britain,' vol. i, part ii, p. 436."

Museum of the Jardin des Plantes, it appears to me that the learned author's description of the specimen, as well as the figure in the plate, are both remarkably correct. Although his specimen does not show the external structure of large *Sigillariæ*, my own observations lead me to the conclusion that we shall find the latter very much resembling, if not altogether identical in structure with, *Sigillaria elegans*. In large specimens of *S. reniformis* and *S. organum*, whose structure is preserved in my own cabinet, there is distinct evidence of the internal cortical envelope, formed of elongated cellular tissue, or utricles, and disposed in radiating series, in all respects like that described by M. Brongniart in his Autun specimen.

"The longitudinal and tangential sections of my specimen show that the vessels of the central axis and the woody cylinder are barred transversely on all their sides. M. Brongniart found this to be the case with *Sigillaria*, and gives it as characteristic of *Sigillaria*, *Stigmaria*, and *Anabathra*.[1] Specimens of these three, now in my cabinet, clearly prove that their central axes and their woody cylinders are exactly the same in structure and arrangement; thus affording evidence, from structure, that *Stigmaria* is the root of *Sigillaria*, and that *Anabathra* is a *Sigillaria*—which has long been expected would prove to be the case."

24. BINNEY[2] (*Diploxylon*):—"This specimen [No. 1] is not in so perfect a state of preservation as those fossil woods intended to be hereinafter described in this communication, especially as regards its central and external parts; but it certainly differs from them in having a larger mass of scalariform tissue composing the central axis, and having the inner portion of the wedge-shaped bundles forming the internal radiating cylinder of a convex shape as they approach the central axis, somewhat like those represented by Brongniart in his *Sigillaria elegans*, and still more resembling those described by Corda in *Diploxylon cycadoïdeum*; but my specimen shows within those convex bundles a broad zone of scalariform tissue, arranged without order, and marked with transverse striæ.

"It has been assumed, both by Corda and Brongniart, that *Diploxylon* had a pith composed of cellular tissue, surrounded by a medullary sheath of hexagonal vessels, arranged without order, barred on all their sides with transverse striæ. My specimen is evidently more complete in structure than those of the last-mentioned authors, or even that which Witham himself described; but, although it shows the so-called medullary sheath in a very perfect state, there is nothing to indicate the former existence of a pith of cellular tissue. All the specimens examined by Witham, Corda, and Brongniart appear to have had their central axes removed altogether, and replaced by mineral matter, or else only showing slight traces of their structure; and these authors appear to have inferred the former existence of a pith of cellular tissue, rather than to have had any direct evidence of it in the specimens of *Anabathra*, *Diploxylon*, and *Sigillaria*

---

[1] " 'Extrait des Archives du Muséum d'Histoire Naturelle,' p. 429, Paris, 1839."
[2] 'Phil. Transactions,' vol. 155, p. 583, 1865.

respectively figured by them. Every collector of Coal-plants is aware of the blank space so generally left in the above fossil plants, as well as in the roots *Stigmariæ*. It is quite true that a little disarrangement of the scalariform vessels (*a′*) in the specimen is seen; but the part which remains undisturbed shows that the whole of the central axis was formerly composed of hexagonal vessels [tubes], arranged without order, having all their sides marked with transverse striæ, and not of cellular tissue. This view is confirmed by another and more perfect specimen of *Anabathra* [*Diploxylon*] in my cabinet, and enables me to speak with positive certainty, and to show that these three plants had a similar structure in the central axes to the specimens of *Sigillaria vascularis* described by me in my paper published in the 'Quarterly Journal of the Geological Society.'

"My specimen clearly proves the existence of medullary rays or bundles traversing the internal woody cylinder, which originate on the outside of the central axis; and it appears to me pretty certain that Corda's specimen of *Diploxylon cycadoideum*, if tangential sections had been made and carefully examined, would have done the same.

"The exterior of the specimen is not in a complete state of preservation, but it seems to have been covered by irregular ribs and furrows, with slight indications of the remains of the cicatrices of leaf-scars. Its marked character, as previously alluded to, is the great space occupied by the central axis. This is of much larger size than in either the *Sigillaria vascularis* or the specimens intended to be next described.

"The lunette-shaped ends of the wedge-like bundles of the inner woody cylinder bear some resemblance to the form of the same parts of the *Sigillaria elegans* of Brongniart; but much more to those of Corda's *Diploxyylon cycadoideum*, with which it appears to be identical. . . . .

"As Brongniart has preferred Corda's name of *Diploxylon* to *Anabathra*, and as the former is a more expressive generic name, in my opinion, probably it is better to adopt it, and accordingly the specimen has been denominated *Diploxylon cycadoideum*."

25. BINNEY[1] (*Sigillaria*):—" Fig. 2 shows the outside appearance of the specimen marked with fine longitudinal striæ, irregular ribs and furrows, and some cicatrices of leaf-scars, which would induce most collectors of Coal-plants to class it with a decorticated specimen of *Sigillaria*. It most resembles *Sigillaria organum*. The bark of a portion of the specimen remains attached to it in the form of coal that is united to the matrix of the seam in which the fossil was found embedded. The reverse side of the specimen does not show the character so distinctly.

"Here we have a *Stigmaria*-like woody cylinder, with a central axis composed of barred vessels arranged without order, found in the inside of a stem of *Sigillaria* in such a position as it existed in the living plant. It is not a solitary instance, but one of more than fifty specimens exhibiting similar characters which have come under my observation.

"In pl. xxxii, fig. 1, is represented the light-coloured disk previously alluded to and

---

[1] Op. cit., p. 586, &c., 1865.

shown in pl. xxxi of the natural size, but here magnified 5 diameters, exhibiting the central axis composed of hexagonal vessels arranged without order, of several sizes, those in the middle being smaller, and becoming larger towards the outside, where they come in contact with the internal radiating cylinder, *b*, and then again diminishing in size. This latter was no doubt cylindrical, like the stem of the plant; but both parts in the process of petrification have been altered by pressure to their present forms. It consists of a broad cylinder, *b*, about an inch in diameter, composed of parallel elongated tetragonal or hexagonal tubes, of equal diameter throughout for the greater part of their length, obtuse and rounded at either extremity, and everywhere marked with crowded parallel lines, which are free or anastomosing all over the surface. The tubes towards the axis are of the smallest diameter; they gradually enlarge towards the circumference, where the largest are situated, though bundles of smaller tubes occasionally occur among the larger. This cylinder, which, for convenience, may be called the internal woody system of the plant, is divided into elongated wedge-shaped masses, pointed at their posterior or inner extremity, and parted by fine medullary rays, of various breadths, some much narrower than the diameter of the tubes, others considerably broader; but none are conspicuous to the naked eye, except towards the outer circumference in some rare instances.

" Fig. 2 represents a transverse section of the central axis and the commencement of the internal radiating cylinder, magnified 12 diameters. The hexagonal vessels in the centre and at the circumference, where they come in contact with the internal radiating cylinder, are smaller in size than those seen in the other parts of the axis. The dark line across the axis, as well as the dark space in the centre, both seem to be the result of a disarrangement of the tubes during the process of mineralization, as similar appearances have not been observed in many other specimens examined by me, which in those parts are in a more perfect state of preservation. The dark and sharp line separating the vessels of the central axis from those of the internal radiating cylinder does not permit us to clearly see the origin of the medullary rays or bundles which undoubtedly traverse the latter.

" Fig. 3 represents a longitudinal section taken on the right-hand side of the specimen, and extending across the whole of the internal radiating cylinder, through the central axis, the intermediate space between the internal radiating cylinder and the outer cylinder, and the external radiating cylinder, to the outside of the stem, magnified 4 diameters; *a, a* showing the smaller barred vessels of the central axis, having some (*a', a'*) which appear to have been disarranged; *b, b* the internal radiating cylinder of larger barred vessels; *c* the space occupied by lax cellular tissue, traversed by bundles of vessels; and *d* the external radiating cylinder, consisting of elongated tubes, or utricles, arranged in radiating series, diverging from certain circular openings, and divided by masses of muriform tissue, which contain the medullary rays or bundles.

" Fig. 4 is a tangential section of the same parts of the specimen as lastly described, magnified 4 diameters ; b', b' showing the medullary rays or bundles traversing the inner radiating cylinder, and d', d' those traversing the outer radiating cylinder.

" Pl. xxxiii, fig. 1, is a longitudinal section of a portion of the same specimen, exhibiting the central axis [1] and the inner radiating cylinder, magnified 15 diameters. Fig. 2 shows several of the vessels of the central axis, as they would be if they were not ground away in the operations of slicing and polishing, magnified 45 diameters. Fig. 3 is a tangential section of the inner radiating cylinder ; b showing the barred vessels, and b'' the medullary rays or bundles, magnified 15 diameters. Figs. 4 and 5, longitudinal and tangential sections of the same specimens, showing the structure of the outer radiating cylinder ; d denoting the tubes, or elongated utricles, of which it is composed, and d' the medullary rays or bundles which traverse it, magnified 10 diameters.

" Pl. xxxvi, fig. 1, represents a transverse section of a ribbed and furrowed stem ('No. 3'), displaying similar cicatrices to that of 'No. 2,' given in pl. xxxi, and having a like central axis, as well as like internal and external radiating cylinders and other parts, magnified 2 diameters. It is given for the purpose of more distinctly showing the tubes or elongated utricles, d, and the fusiform openings formed of very open muriform tissue, d', enclosing the medullary rays or bundles which traverse the external radiating cylinder ; this it does in a very marked manner ; magnified 20 diameters.

" In pl. xxxv, figs. 1, 2, and 3 ('Nos. 4, 5, and 6'), are shown the exteriors of three central axes, separated from large-ribbed-and-furrowed stems, in every respect similar to those described in pl. xxxi, and pl. xxxiv ; and such as might easily be taken for small *Calamites;* magnified $2\frac{1}{2}$ diameters. Fig 4 ('No. 7') shows the outside of the internal woody cylinder of a *Stigmaria*, with ribbed and furrowed characters, resembling those shown on the outsides of the central axes lastly described ; also magnified $2\frac{1}{2}$ diameters. The first three specimens, 'Nos. 4, 5, and 6,' are from the Halifax 'Hard Seam ' of coal, at South Owram ; but 'No. 7' is from the Wigan " Five-feet Mine," a seam in the Middle Coal-measures. The tangential sections which show the medullary rays, or bundles that traverse the inner and outer radiating cylinders, afford clear evidence of the different appearance of the bundles marked b'' in pl. xxxiii, fig. 3, from those in pl. xxxiv, fig. 2, marked d'.

" Specimens 'Nos. 2 and 3' bear considerable resemblance to the *Sigillaria elegans* of Brongniart, with respect to their internal radiating cylinder and the medullary rays or bundles which traverse it, assuming that such vessels come from the outside of the central axis, and not from the exterior of the internal radiating cylinder, as that distinguished savant supposed. Certainly there is no evidence in my specimens to support the latter view. A great many specimens have been broken up and destroyed for the

---

[1] "In the plate the small tubes a', a'' appear to be divided by septa. This is certainly the case in one slice ; but in another of the same specimen these septa are not seen, but small barred vessels appear in their places, so the former may probably be due to the direction of the slice being cut along the dark line which traverses the central axis, as shown in pl. xxxii, figs. 1 and 2."

purpose of examining the inner radiating cylinder, and in every case medullary rays or bundles were found traversing it, just as you find in the same part of *Stigmaria*. On the outside of the inner cylinder, at the extreme part of the zone of coarse and lax cellular tissue which bounds it, are some circular openings from which spring the wedge-shaped masses of quadrangular, tubular, or elongated utricles which form the outer radiating cylinder. The lax cellular tissue has nearly always been displaced and disarranged in the process of mineralization, and sometimes the outer radiating cylinder, and the circular orifices connected with it, have been pushed towards the inner cylinder. This may be the case in Brongniart's specimen, and have caused him to suppose that the medullary rays or bundles originated only on the outside, and were not joined to those which traversed the inner cylinder. So far as my larger specimens show there were medullary rays, which had their origin next the central axis, passed through the inner cylinder, and, after traversing the zone of lax cellular tissue outside the latter, apparently communicated with similar rays or bundles of vessels of much larger size, which are always found traversing the outer radiating cylinder, and then went on to the leaves on the outside of the stem. . . . . . .

"In the specimens Nos. 2 and 3 the outer radiating cylinders are nearly an inch and a half in breadth, of thick-walled tubes or elongated utricles arranged in radiating series and diverging from a circular opening; while in Brongniart's *Sigillaria elegans* the outer radiating cylinder was not more than $\frac{1}{12}$th of that breadth. Probably my specimens may not prove to be of the same species as that of the celebrated Autun specimen; still they may be of the same genus, although of considerably greater age. But they have the greatest resemblance to the *Sigillaria vascularis* described by me in a paper read before the Geological Society and printed in its 'Journal' (vol. xv, p. 636). All the specimens described in that communication, as well as those in the present, were obtained by me from the same seam of coal, but at different places; still the two, namely the large-ribbed-and-furrowed specimens and the small rhomboidal-scarred stems, are always found associated together, and they can be traced gradually passing from one into the other. These facts, when taken in connection with the similarity in structure in the central axis, the internal radiating cylinder, the space filled with lax cellular tissue between the latter and the outer radiating cylinder diverging from circular openings, clearly prove that the smaller specimen is but the young branch of the older stem 'No. 2.' It is true that the earlier authors who have written on these plants would scarcely have admitted a ribbed and furrowed *Sigillaria* to have been so intimately connected with a rhomboidal-scarred plant, but it is now generally allowed that such differences in external characters would afford no grounds for ignoring the structural similarity of the specimens. Undoubtedly the small *Sigillaria vascularis* was part of a branching stem, for in my cabinet there is a specimen clearly showing two internal radiating cylinders just at the point where the branches dichotomized, as shown in the woodcut (fig. 2), so often met with in *Lepidodendron*.[1] . . . . . .

[1] See Plate XIV, fig. 4, of this Monograph.

" The broad space intervening between the internal and external radiating cylinders, filled with lax cellular tissue and traversed by medullary bundles communicating with the leaves on the outside of the stem, as shown in the specimens described in this paper, is the only part on which information is required to complete our knowledge of the structure of the stem of *Sigillaria*. Fortunately a small specimen of *Sigillaria vascularis*, kindly presented to me by Mr. Ward, of Longton, a most indefatigable collector, has enabled me to obtain considerable information on this point. This specimen shows the rhomboidal scars on the outside of the stem, the two radiating cylinders, and the space between occupied by lax cellular tissue and traversed by medullary bundles.

FIG. 4.

Sigillaria vascularis.

" The specimen in this woodcut [1] (fig. 5 [fig. 4], magnified twice) is of smaller size than any previously described by me, but it is, from both its internal structure and external characters, a small *Sigillaria vascularis* in its young state, when the two radiating cylinders, especially the outer one, of the plant were only slightly developed. The medullary rays are seen on the outside of the inner radiating cylinder, and pass, inclining upwards at a small angle, from the inner cylinder to nearly the outside of the stem. No trace of the outer cylinder can be seen, so as to enable us to see whether the smaller-sized medullary bundles, coming from the inner cylinder, join the larger ones in the outer cylinder, described in pl. xxxiv, fig. 2, and there marked *d'*. All the tangential sections show the medullary bundles, both in large and small specimens, to be much greater and stronger in the outer than in the inner radiating cylinder; but no evidence has yet been found of the junction of these medullary bundles to prove that the former run into the latter, or whether the two are distinct. They consist of hexagonal tubes, barred on all their sides, surrounded by muriform tissue, that on the outside of the specimen being of very coarse texture."[2]

[1] Obligingly lent by the Council of the Royal Society.

[2] In all the large and small specimens of *Sigillaria vascularis* which have come under my observation that illustrated by this woodcut is the only one that clearly shows the vascular or foliar bundles proceeding direct from the outside of the inner radiating cylinder to the leaf-scars. This, from recent investigations, has been known to be the case. On its outside it is covered with rhomboidal scars like all the small specimens. The space intervening between the inner radiating cylinder and the outer one appears to have once consisted of iron-pyrites, which has since been decomposed, leaving the vascular or foliar bundles fully exposed. On comparing the direction which these organs take, from the inner to the outer radiating cylinder, with those shown in the specimen "No. 31," Plate XIII, figs. 2 and 3, of *Lepidodendron Harcourtii* in this Monograph, it will be seen that they run in nearly a horizontal direction, compared with the high angle the latter make when proceeding from the stem. This difference in the direction of the vascular or foliar bundles in *Lepidodendron* and *Sigillaria vascularis* is very marked, and worthy of the attention of those authors who contend that the latter plant is only a *Lepidodendron*.—E. W. B.

26. Binney[1] (*Stigmaria*) :—"Many years since, after an examination of a great number of specimens of *Stigmaria* in my collection, it occurred to me that an outer radiating cylinder would ultimately be discovered. In my remarks on *Stigmaria*[2] is the following passage :—' That part of *Stigmaria* which intervened between the vascular axis and the bark appears to have consisted of two different kinds of cellular tissue. These in most cases have been unfortunately destroyed, so that we cannot positively know their true nature ; but they appear to be of different characters, for there generally appears to be a well-marked division. This is often shown in specimens composed of clay-ironstone which have not been flattened, and, the boundary-line is generally about a quarter of an inch from the outside of the specimen. More probably the outer part of the zone has been composed of stronger tissue than the inner one, as is the case with well-preserved specimens of *Lepidodendron*.' It is singular that Drs. Lindley and Hooker, as well as such acute observers as Brongniart and Göppert, had not noticed this line of division ; but this was, no doubt, owing to the imperfect specimens they had examined. After the discovery of the outer radiating cylinder by Witham in *Lepidodendron*, and the same arrangement in *Sigillaria* by Brongniart, it was to be expected that such outer radiating cylinder would be found to occur in *Stigmaria*, if it were the root of *Sigillaria*. After an inspection of a great number of specimens, the cabinet of Mr. Russell, of Chapel Hall, Airdrie, has afforded me four or five distinct specimens which give clear evidence of the existence of this outer radiating cylinder in *Stigmaria*. They are all in clay-ironstone, and have not been much compressed. He has kindly allowed me to slice two of the specimens, which afford decisive evidence of the former existence of both an inner and an outer radiating cylinder. The space on the outside of the inner cylinder does not distinctly show the bundles of vessels communicating with the rootlets, although there is some evidence of their former existence. The bell-shaped orifices from which the rootlets spring are well displayed, and the space between them is occupied by wedge-shaped masses of tubes or elongated utricles arranged in radiating series, and not to be distinguished in any way from those shown in pl. xxxv, fig. 5. Indeed, the transverse section of the specimens there figured

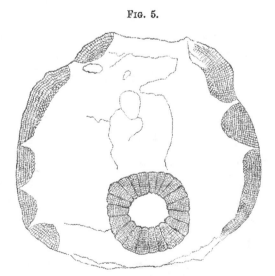

Fig. 5.

Section of *Stigmaria*, from Airdrie.

[1] 'Phil. Trans.,' vol. clv, 1865, p. 592.

[2] 'Quarterly Journal of the Geological Society,' vol. vi, p. 20.

[3] For this and the preceding woodcut I am indebted to the Council of the Royal Society, who have kindly lent me the blocks.

would almost do for a representation of the *Stigmaria*, if the latter had the central axis preserved, which it unfortunately has not. There is the same internal radiating cylinder, the same space occupied by lax cellular tissue, which gradually passes into tubes or elongated utricles, arranged in radiating series, apparently diverging from circular openings, and parted by large bundles of muriform tissue containing vessels barred on all their sides, extending to the outer bark. The accompanying woodcut (fig. 5) will give a much better idea of its structure than our laboured description.

"This specimen clearly proves, by the evidence of internal structure alone, that *Stigmaria* is the root of *Sigillaria*, each of them having an inner radiating cylinder composed of barred vessels, a space occupied by lax cellular tissue, and an outer radiating cylinder composed of tubes or elongated utricles.

27. CARRUTHERS (*Sigillaria, &c.*, 1866).—For the views of this author see page 65 of this Monograph.

28. SCHIMPER (*Sigillaria*).—Professor Schimper[1] classes *Sigillaria* under the *Lycopodinées*, and in the family *Sigillariæ*. He writes—"Trunci cylindrici, simplices vel apice pluries furcati, longitudinaliter sulcati vel læves, foliorum cicatricibus regularibus spiraliter dispositis ornati, cylindro axili continuo vel radiis medullaribus (fasciculis vascularibus?) pertuso medullam crassum includente instructi, cæterum e parenchymate (vivo succulento) cortice solido tecto compositi. Radices crassæ, pluries dichotomæ, longissimæ, horizontaliter expansæ, radiculis longis, simplicibus, crassiusculis, spiraliter dispositis, articulatione circulari insertis. Folia graminiformia triplicata nervo simplici percursa, post lapsum cicatrices relinquentia ovatas, ovato-hexagonas, exacte hexagonas, vel transverse rhombeas, vasorum cicatriculis tribus notatas, medio punctiformi, duabus lateralibus semilunaribus; cicatriculis trunci decortati binis, approximatis hic illic in unam confluentibus, ovalibusve linearibus. Fructificatio spicæformis, sporangiis bractearum basi dilatata insertis.

"D'accord avec la plupart des auteurs modernes, je range les Sigillariées dans l'ordre des Lycopodiacées, malgré la présence des rayons médullaires dans le cylindre ligneux, dont, d'après M. Brongniart, les ' vaisseaux rayés et réticulés' seraient ' disposés en séries rayonnantes' comme dans les Cycadées, ce qui a engagé ce savant à réunir ces plantes aux Gymnospermes. La nature des vaisseaux, en grande partie scalariformes, le vaste parenchyme qui recouvre le cylindre ligneux, la forme régulière des cicatrices foliaires et celles des feuilles elles-mêmes, enfin la mode de fructification, qui est celle des Lycopodiacées, sont des caractères qui rapprochent ces singuliers fossiles plutôt des Lépidodendrées que de tout autre type végétal. D'après plusieurs observations récentes, la végétation souterraine des *Lépidodendron* aurait même une grande ressemblance avec celle des *Sigillaria*. Cette végétation était formée par des racines puissantes, ramifiées par dichotomie répétée, s'étendant horizontalement à de grandes distances et garnies de

---

[1] 'Traité de Paléontologie Végétale,' tome ii, première partie, p. 76, 1870.

radicelles épaisses, charnues, disposées en spirale, et se désarticulant, comme les feuilles, en laissant des cicatrices persistantes circulaires. Les Sigillariées comptent parmi les plantes les plus communes dans le terrain houiller, et paraissent avoir habité de préférence les endroits marécageux.

"A. *Trunci, Sigillaria*, Brong., *Syringodendron*, Sternb., Brongt., ex. p. Atlas, pl. lxvii, lxviii.

"*Trunci arborei, elati, crassi, simplices rarius, ad apicem dichotomi.* Foliorum cicatrices, rectiseriatæ, seriebus sulco a se invicem separatis, vel contiguæ corticemque clathrato-reticulatum reddentes, vel tandem distantes atque cortici lævi vel leniter ruguloso insidentes, nunc ovales apiceque truncatæ vel emarginatæ, nunc ovato seu regulariter hexagonæ, rarius transverse rhombeæ diagonali transversa longiore quam recta; cicatriculis fasciculorum vascularium tribus, medio punctiformi, lataralibus lunularibus. Folia ipsa linearia, longa, subplana, vel triplicata, plicis carinatis, spiraliter vel verticillatim disposita.

"Les troncs des *Sigillaria* peuvent être divisés en deux groupes, en troncs cannelés et en troncs lisses. Les premiers sont parcourus de côtes aplaties verticales parallèles, dont chacune porte une seule série de cicatrices; ces côtes ont leurs côtés exactement parallèles, ou elles sont plus ou moins distinctement étranglées entre les cicatrices. Dans les formes où ces côtes n'existent pas les cicatrices sont contiguës, et recouvrent toute la surface du tronc, ou elles sont séparées par des espaces lisses plus ou moins large. Après la chute de l'écorce il ne reste plus sur le tronc que les cicatricules des faisceaux vasculaires, très-variables quant à leur grandeur, ovalaires, réunies ensemble ou confondues en une seule, saillantes ou enfoncées dans une fossette (pl. 67, f. 8, 9, pl. 58). L'arrangement phyllotaxique des cicatrices est analogue à celui des *Lepidodendron*.[1] On remarque assez souvent, entre les séries des cicatrices foliaires, des séries interrompues de cicatrices tout-à-fait différentes de ces dernières. Ces cicatrices sont ovalaires, convexes, ombiliquées au centre, d'où partent en rayonnant plusieurs rides (voy. pl. 67, f. 2 *a*). Ce sont probablement les cicatrices d'insertion des épis fertiles (figs. 12, 13, 14). Sur une espèce, le *Sig. spinulosa*, espèce, dont les cicatrices foliaires sont espacées, et l'écorce lisse, il existe immédiatement sous ces dernières une ou deux petites cicatrices circulaires à bord relevé en bourrelet et ombiliquées au centre. Ces cicatrices ont été prises pour des cicatrices provenant d'épines dont cette espèce aurait été munie, à l'instar de quelques Euphorbiacées frutescentes ou arborescentes. J'y vois des cicatrices de racines adventives. Leur forme est en petit celle de cicatrices des *Stigmaria* (voy. figs. 12, 12 *b*).

"Comme dans les *Lepidodendron*, la structure microscopique du tronc n'a encore été reconnue que sur un très-petit nombre de fragments silicifiés. Je n'ai jamais eu occasion d'examiner de pareils fragments en détail, et me vois, par conséquent, obligé de m'en rapporter à ce qui a été publié sur ce sujet. M. Brongniart, qui a été assez heureux de

---

[1] Voyez, pour la disposition des feuilles dans les *Sigillaria*, Naumann, 'Ueber d. Quincunx als Gesetz der Blattstellung vieler Pflanzen,' Leipzig, 1845. Goldenberg, 'Flora Sarœp. Foss.,' livre 2, p. 1, et suiv.

pouvoir étudier un échantillon silicifié du *Sigillaria elegans*,[1] dit dans son 'Tableau des genres de végétaux fossiles,' p. 55 — 'Le caractère essentiel de ces plantes c'est de présenter, dans l'intérieur de leur tige, un cylindre ligneux entièrement composé de vaisseaux rayés ou réticulés, disposés en séries rayonnantes, séparés en général par des rayons médullaires ou par les faisceaux vasculaires qui, de l'étui médullaire, se portent vers les feuilles. Cette organisation est presque identique avec celle des Cycadées; mais outre la différence des formes extérieures, les principaux genres de cette famille, ceux qui appartiennent sans aucun doute à de vraies tiges, présentent, en dedans du cylindre ligneux dont je viens de parler, un cylindre intérieur, sorte d'étui médullaire, continu et sans rayons médullaires dans le *Diploxylon*, divisé en faisceaux correspondant aux faisceaux principaux du cylindre ligneux dans le *Sigillaria*.' Je ne pense pas qu'on puisse prendre les lames parenchymateuses qui séparent les faisceaux vasculaires dont se compose le cylindre ligneux pour des rayons médullaires dans le sens propre du mot. Nous voyons aussi dans d'autres Lycopodiacées les faisceaux vasculaires qui concourent à la formation du cylindre ligneux séparés les uns des autres par un tissu parenchymateux qui se confond avec le tissu médullaire central. M. Binney[2] décrit et figure le cylindre ligneux intérieur de son *Sigillaria vascularis* comme entièrement occupé par un tissu composé de larges vaisseaux scalariformes et d'autres de moindres dimensions (vaisseaux spirals). Le même auteur dit que les rayons médullaires qui passent entre les faisceaux vasculaires dont se compose le cylindre ligneux extérieur sont formés par des vaisseaux finement rayés et se rendent dans les feuilles ! Nous aurions donc affaire plutôt à des faisceaux vasculaires partant de l'intérieur du cylindre ligneux que des rayons médullaires. Cela paraît être mis hors de doute par la figure qu'a publiée M. Binney (' Phil. Transact.,' *l. c.*, p. 594) d'un fragment de jeune tige de *S. vascularis*, dans lequel ces soi-disant rayons médullaires sont régulièrement disposés en quinconces. Le cylindre ligneux extérieur est suivi d'un large parenchyme à cellules très-délicates, auquel succède un tissu cellulaire plus lâche, limité extérieurement par le tissu cortical, très-serré, et solide.

"Dans le *S. elegans*, Brong., le cylindre ligneux est formé en partie vers sa partie intérieure de vaisseaux spirals très-étroits ; ce même genre de vaisseaux se trouve aussi dans le cylindre médullaire (*voy.* Brong., *l. c.*, pl. xxviii, f. 1 *b, b*, B).

"J'ai déjà fait remarquer plus haut que les cicatrices foliaires du *Sigillaria vascularis*, Binney, ne diffèrent pas de celles du *Lepidodendron vasculaire* du même auteur ; j'ajouterai

[1] Brongniart, "Observations sur la Structure Intérieure du *Sigillaria elegans*," &c., ' Arch. d. Mus. d'Hist. Nat.,' tome i, p. 406, pl. xxv–xxviii.

[2] E. W. Binney, "On some Fossil Plants showing Structure, *Sigillaria* and *Lepidodendron*," ' Quart. Journ. of the Geol. Soc.,' May, 1862, p. 106, pl. iv, v ; *idem*, ' Philosoph. Transact.,' mdccclxv, p. 580, pl. xxxi–xxxv. M. Binney figure à la pl. xxxv, f. 6, un échantillon qu'il rapporte au *S. vascularis*, et dont les cicatrices foliaires dénotent évidemment un *Lepidodendron* très-voisin du *L. Veltheimianum*, et probablement identique avec son *S. vasculare*. Voyez aussi, Dawson, "On Vegetable Structure of Coal," ' Quart. Journ. of the Geol. Soc.,' vol. xv, p. 636 ; *idem*, "Coal. Format. of N. Scotia and New Brunswick," ' Quart. Journ. Geo. Soc.,' xxi, 1865.

encore que la surface extérieure du tronc muni de ces cicatrices, que M. Binney a figuré à la pl. xxxv, f. 6, de son ' Descript. of some Fossil Plants showing Structure ' (' Philos. Trans.,' vol. mdccclxv), ressemble à un tel point à celle du *Sagenaria fusiformis*, Corda, (' Beitr.,' tab. vi), qu'il est impossible de l'en distinguer. Ce *Sagenaria* est très-voisin du type de *Lepidodendron* représenté par le *L. Veltheimianum*, type qui pourrait bien former le passage de ce genre au genre *Sigillaria*.

" De nombreuses observations paraissent prouver à l'évidence que le *Lepid. Veltheimianum* possédait pour racine ou rhizome un *Stigmaria ;* nous aurions là une nouvelle preuve pour ce passage.

" Il résulte de tout ce que nous venons de dire que, malgré les beaux travaux qui ont été faits sur ce sujet, notre connaissance sur la structure microscopique des tiges de *Sigillaria* laisse encore beaucoup à désirer. Mais je crois que M. Binney a parfaitement raison quand il dit, ' Everything has led me to believe that the leaves and branches (?), and probably the fructification of *Sigillaria* would prove to be very analogous to those of *Lepidodendron* ' (*loc. cit.*, p. 591).'

" Les *Sigillaria* n'ont jamais été rencontrés en dehors du terrain houiller, et ils abondent surtout dans les formations houillères moyennes et supérieures, dans lesquelles on a souvent observé des troncs d'une hauteur considérable, occupant encore leur position verticale primitive, mais ne montrant jamais aucune ramification. C'est ainsi qu'on a découvert, en construisant le chemin-de-fer de Saarbrücken à Neunkirchen, toute une forêt de Sigillaires encore debout. Dawson a vu la même chose dans les houillères de la Nouvelle Écosse. À Saint-Étienne et à Anzice, en France, les troncs de Sigillaires traversant perpendiculairement plusieurs couches houillères ne sont pas rares. En Europe, comme en Amérique, ce sont surtout les troncs des *S. reniformis* et *lævigata* qui ont conservé ainsi leur position primitive."

29. WILLIAMSON (*Stigmaria*, 1871).—See page 71 of this Monograph.

30. WILLIAMSON (*Diploxylon*).—After adopting Mr. Carruther's views as to some of the specimens of my *Stigmaria vascularis* belonging to *Lepidodendra*, Professor Williamson[2] treats of *Diploxylon*. He states, " My specimens throw no direct light upon the structure of the vascular and medullary axis of the true *Sigillariæ* as distinguished from the Favularian type; but the cortical portions of all the plants, including the true *Sigillariæ*, exhibit what is practically an identity of structure. In all we have a remarkably thick, spongy bark, reminding us, in many of its features, of that found in the living Cycads. This consisted either of parenchyma, prosenchyma, or of both combined, enclosed externally in a vast layer of elongated prosenchymatous tubes, which, in turn, is invested by a layer of cellular parenchyma supporting the bases of leaves, the latter invariably consisting of the same form of parenchyma as the epidermis. M. Brongniart's specimen of *Sigillaria (Favularia) elegans* exhibits a central axis the structure of which is nearly iden-

[1] As to *Knorria* having Stigmaroid roots, see p. 89 of this Monograph.—E. W. B.
[2] ' Philosophical Transactions' for 1872, p. 198, and p. 227.

tical with that of my specimen (pl. xxviii, figs. 33 and 34). This, in its turn, only differs from the more ordinary forms of *Diploxylon* in the crenulated outline which separates the ligneous zone from the cylinder of medullary vessels, giving to the exterior of the latter a fluted aspect, like that of a Calamite, but without the transverse nodal constriction of the latter genus. The Diploxylons, again, as I have already shown, shade off into the ordinary forms of *Lepidodendron*, and are, undoubtedly, Lepidodendroid plants which have lost the central portion of their medullary axis. Remove the cellular tissues from the centre of the plant which I have represented in figs. 8 and 9, and we have at once the closest resemblance to Witham's *Anabathra* and Corda's *Diploxylon*, as well as to those now under consideration. That Witham's plant is identical in type with mine is further indicated by his tab. 8, fig. 12, where he exhibits one of the large compound medullary rays shown in my pl. xxvii, fig. 23. The cellular tissues have not been preserved in the medullary rays of Brongniart's *Sigillaria elegans*; but tab. 4, fig. 2, of his memoir shows that his plant possessed similar ones to those which Witham and I have figured. Further, the description which M. Brongniart has given of the *outer* bark and epidermis of his plant, these being the only cortical elements remaining in his specimen, would apply, with little or no alteration, to several of my Lepidodendroid and Sigillarian types; so that, whilst a really indisputable *Sigillaria*, like my pl. xxix, fig. 39, but in which the woody axis is preserved *in situ*, is still an important desideratum, I have very little doubt that, when discovered, it will be found to correspond with one of the several varieties of *Diploxylon*. Most probably, also, my pl. xxv, fig. 8, representing one of the extremes of the two types figured by Mr. Binney under the name of *Sigillaria vascularis*, will also be found to belong to the same subtype of the same genus. Yet my indefatigable friend informs me that his cabinet contains specimens in which the most gradual transition can be traced, from the plant just referred to, to the *Lepidodendron selaginoides*, the oppositely divergent form of the same group, hence his inclusion of both under one common name.

. . . . " Having thus obtained (WILLIAMSON, *op. cit.*, p. 238) additional light respecting the Diploxylons, I again turned to the more highly organised of the stems described by Mr. Binney under the name of *Sigillaria vascularis*, and which I have already represented in pl. xxv, figs. 8—11. I made a fresh series of carefully prepared dissections, and succeeded in demonstrating the existence, in this plant, of a series of primary and secondary medullary rays, the former containing large foliar bundles precisely identical with those of *Diploxylon cycadoideum*. I have not succeeded in discovering in the former plant the cellular layer intervening between the medullary vascular cylinder and the woody zone of the latter one. The large primary medullary rays are composed of barred cells, which are sometimes mural, but more frequently prosenchymatous; through the upper part of each of these large rays there proceeds a bundle of true barred vessels. I have not succeeded in tracing one of these bundles to its medullary extremity, consequently I cannot yet affirm how it originates; but I have seen sufficient to confirm what I have already stated in the body of the memoir, that we need only remove the central cellular medulla of the plant in question to convert it into a true *Diploxylon*; the identity of the two, so far as structural type is concerned, is as close as it can be even in its minuter details. Such being my conviction, I propose to designate the plant represented in figs. 8—11 *Diploxylon vasculare*, and to apply Corda's name of *D. cycadoideum* to figs. 21—23.

"The plant represented by figs. 33, 34 distinguished by its large medullary axis and by the deeply fluted aspect of the interior surface of its ligneous zone, I propose to designate *Diploxylon cylindricum*; whilst a fourth form, exhibiting some different features yet to be noticed, I would term *D. stigmarioideum*. So far as the general structure of the stem is concerned, the last-named plant does not differ from the other Diploxylons. The cellular medulla has disappeared, but there remains the medullary ring of barred vessels, surrounded by the exogenous ligneous zone. The primary and secondary medullary rays also appear, but neither of them occurs so abundantly as in the other species. Moreover, in the radial vertical sections, the vascular bundles occupying the primary rays exhibit a different aspect from those of the other species described, and approach nearer to what exists in *Stigmaria ficoides*. This is represented in fig. 23, *b*. The vascular bundle (*m*) appears to be derived from the body of the ligneous zone, and not from its medullary surface. It is composed of smaller vessels than those seen at *e*; but we find that at *e'* these vessels diminish in size, and approach in magnitude those of the bundle *m*; not only so, but, whilst the upper extremities of the small vessels of *m* exhibit the perpendicular arrangement indicating that they belong to the part of the woody zone in which they occur, the lower extremities of the large vessels (*e*) are deflected in the direction of those of the foliar bundle, which is never the case with the corresponding ones of the other forms of Diploxylons. The lower margin of the foliar bundle is cut off in this section by an oblique, sharply defined line; this indicates that the large vessels at *e″* have been sharply deflected to the right and left of the bundle, to allow the latter to pass between them. All these appearances correspond so closely with what we find in *Stigmaria* that for a long time this plant seriously perplexed me; but it appears to be a true *Diploxylon*, since it has the vascular medullary cylinder of that genus as well defined as in any other species. This cylinder is never found in *Stigmaria ficoides*. It has been more especially in connection with this species of *Diploxylon*, though not exclusively, that I have found the peculiar bark represented in figs. 54—57. It is possible that this plant may, like *Stigmaria*, prove to be the uppermost part of a root of some of the other forms, though I have never yet found it associated with any rootlets; or it may be a fragment from the base where stem and roots united.

"Amongst the numerous other interesting plants for which I am indebted to G. Grieve, Esq., of Burntisland, in Fifeshire, is a well-marked *Diyloxylon*, closely allied to *D. cyca-deoideum*. Like the rest of Mr. Grieve's specimens, it is from the deposit of the Lower Carboniferous age which occurs embedded amongst trappean rocks at Pettycur Bay.

This specimen is instructive, since, though abundantly furnished with primary and secondary medullary rays, or rather with the spaces which they occupied, all the cellular tissues have disappeared from both, whilst the vascular foliar bundles are well preserved. We are thus enabled to distinguish the respective areas occupied by the two tissues in a manner that I have not succeeded in doing so distinctly in the other specimens described. Each bundle is cylindrical, occupying the centre of the lenticular section of the ray, when cut at right angles to its direction, and consisting of very small, barred vessels. Above and below the vessels are open spaces; but these were originally occupied by the cellular tissues of the ray, the forms of the cells being strongly impresssed upon the indented walls of the contiguous longitudinal vessels of the ligneous zone. I have not discovered in this plant the cellular layer intervening between the medullary vascular cylinder and the woody zone; in this respect it appears to approach nearer to *D. vasculare* than to the other forms. The vascular medullary cylinder or sheath is strangely marked; but all the medullary cellular tissues have disappeared. I pointed out some time ago[1] that some of these *Lepidodendra* exhibited a feature not previously noticed; namely, the vessels were not only barred transversely, but, in addition, the transverse bars of lignine were connected by a delicate series of threads of the same material, running parallel with the longer axis of the vessel. I find this feature in all Diploxylons; but in the Burntisland specimen it is so faint that it can only be discovered under the microscope by a careful adjustment of the light. The coarser transverse bars are also much more irregular in size, number, and direction than is usual amongst the Diploxylons of the Upper Coal-measures.

"The *Diploxylon* of Corda is so obviously identical generically with the *Anabathra* of Witham that the latter name ought to be adopted in preference to the former one. But ere long, in all probability, both these names will have to be abandoned, since there appears little doubt that they represent the woody axes of some of the common Lepidodendroid plants of the Coal-measures; and, as soon as the identification of these internal axes with their correlate external forms is indisputably accomplished, the yet older names of the latter must become the adopted ones. Under these circumstances it is scarcely desirable to disturb a widely accepted nomenclature, since any day may furnish the required connecting link.

"The general conclusion towards which all these additional observations point is the same as that of the preceding memoir, which they strengthen and confirm, viz. that all these varied plants are constructed upon a common type, and belong to one Lycopodiaceous family."

31. NEWBERRY[2] (*Sigillaria*):—"Fossil-botanists have discussed the relations of *Sigillaria* at considerable length, without reaching any universally accepted conclusion. Professor Dawson considers they are Gymnosperms, while Mr. Carruthers regards them as distinctly Cryptogamous, and more nearly allied to Lycopods than to the Conifers. My

[1] 'Monthly Microscopical Journal,' August 1st, 1869, pl. xx, fig. 10.
[2] "Report of the Geological Survey of Ohio," vol. i, Part II, 'Palæontology,' p. B, 65, 1873.

own observations confirm those of Professor Dawson in regard to the structure of the trunk of *Sigillaria*. The outside was evidently composed of a thick cortical integument, to which the leaves were attached. Within this was a mass of cellular tissue, surrounded by a slender woody axis, with a relatively large medullary cavity. This is very unlike the structure of the trunk in most of our Conifers, but it is not very dissimilar to the trunk of Cycads. The probabilities are, that the *Sigillariæ* formed a group of plants considerably unlike any now living, and, as such, served to connect the Gymnosperms with the Acrogens. If this was their botanical position, it would not be at all surprising if we found that they possessed a trunk sharing the peculiarities of the Sago-palms and Tree-ferns, bearing drupaceous fruits not unlike those of the Cycads and some of the infinitely varied Coniferæ. If we compare the fruits of *Pinus*, *Taxus*, *Salisburia*, and *Ephodia*, among the *Coniferæ*, we shall discover such a latitude of structure as will prepare us to accept the association of the fruit of *Trigonocarpon* with the trunk of *Sigillaria* without much hesitation."

32. RENAULT AND GRAND' EURY[1] (*Sigillaria spinulosa*) :—"Des faits qui précèdent il résulte principalement que les vraies Sigillaires, ainsi que le pensait déjà M. Brongniart, ont les éléments ligneux arrangés en séries radiales et croisantes, séparés par de vrais rayons médullaires, comme les plantes phanérogames gymnospermes; que les faisceaux foliaires tirent leur origine de l'étui médullaire, comme il arrive chez les plantes dicotylédones. Entre le cylindre ligneux et l'étui médullaire il n'existe aucune couche cellulaire analogue à celle du *Diploxylon cycadeoidum* signalée par M. Williamson. Les cellules des rayons médullaires ne sont pas barrées comme celles qui forment les rayons médullaires du *Diploxylon* et du *S. vascularis*. Les faisceaux foliaires partent de la portion intérieure et médiane des faisceaux médullaires, celle qui est composée de vaisseaux plus petits, barrés et spirals, et, après avoir traversé le bois obliquement, ils s'élèvent verticalement dans la zone parenchymateuse de l'écorce, et s'infléchissent ensuite pour parcourir presque horizontalement la partie tubéreuse. De chaque côté du faisceau foliaire deux lacunes, parcourues par des canaux volumineux, prennent leur origine dans le tissu cellulaire sous-cortical, et viennent former à l'extérieur, sur la cicatrice, ces deux arcs placés de chaque côté du faisceau foliaire, médian et unique, et si apparents dans les Sigillaires. L'écorce tubéreuse est parcourue obliquement, de bas en haut, par de nombreux rayons cellulaires, limités par un tissu formé de cellules extrêmement régulières, disposées par bandes rayonnantes. Par les caractères les plus essentiels, les *Sigillaires* ont donc bien l'organisation des tiges dicotylédonées, et particulièrement des Gymnospermes et surtout des Cycadées.

"Dans le mémoire cité plus haut M. Williamson, après avoir fait ressortir les analogies existant entre quelques portions de Lépidodendrées, le *Sigillaria vascularis*, les *Diploxylon* et les vraies *Sigillaires*, dont il ne met pas en doute les caractères phanéro-

---

[1] 'Mémoires à l'Académie des Sciences, &c.,' vol. xxii, No. 9, p. 16, 1874.

gamiques, qui vont, au contraire, comme il le fait marquer, en s'accusant de plus en plus dans ces dernières, conclut que toutes ces variétés de plantes ont le même prototype, et qu'elles appartiennent à une même famille, les Lycopodiacées.

"Tout en reconnaissant ce qu'il y a de séduisant à admettre l'existence d'une longue série de plantes cryptogamiques de plus en plus élevées en organisation, et dont l'un des termes les plus parfaits serait les vraies Sigillaires, qui offrent dans leur structure les principaux traits des Phanérogames—tout en admettant l'importance philosophique d'une hypothèse qui n'a rien que de très-naturel et de très-conforme à ce qui existe dans d'autres branches de l'histoire des êtres—nous pensons devoir rester dans une sage réserve, en attendant le moment, peut-être prochain, où des fructifications parfaitement authentiques et bien conservées permettront de trancher définitivement la question."

33. BRONGNIART.[1]—In writing on *Trigonocarpon* M. Brongniart states in a note— "Je dois rappeler ici que j'ai toujours considéré, d'après la structure de leurs tiges, les *Sigillaria* et les *Calamodendron* comme se reportant à des types détruits de végétaux arborescents de la grande division des Dicotylédones Gymnospermes, contrairement à l'opinion de plusieurs paléontologistes, qui les rangent parmi les Cryptogames, près des Lycopodiacées et des Equisétacées.

"Le nombre et la variété des graines de Gymnospermes que je signale dans ce memoire confirment cette opinion, que je vois avec satisfaction adoptée par M. Newberry dans son mémoire récent sur diverses graines des terrains houillers de l'État de l'Ohio."

IV. DESCRIPTION OF THE SPECIMENS.

§ 1. SPECIMENS Nos. 39 and 40, *Sigillaria vascularis*, Binney. Pl. XIX, figs. 1 and 2; Pl. XX, figs. 1, 2, 3, 4, and 5.

The first specimen intended to be described in this Memoir is from a calcareous nodule found in the Halifax "Hard Seam" of coal at South Owram, Yorkshire, and marked "**" in the section of strata given at page 12. It was associated with *Sigillaria, Stigmaria, Lepidodendron, Calamodendron*, and other Coal-measure Plants. The chemical composition of the nodule of limestone and the circumstances connected with the occurrence of the fossil wood are the same as those previously described with regard to No. 3 specimen in this Monograph. The specimen illustrated in Plate XIX, figs. 1 and 2, No. 39, natural size, is of an irregular oval shape, one foot in circumference, five inches in its major and three inches in its minor diameter. When first discovered it was six inches in length, being a fragment of a much larger stem. The light-coloured disk in the middle, about an inch in diameter, shows the central axis and the internal

---

[1] 'Extrait des Annales des Sciences Naturelles,' "Botanique," 5e série, vol. xx, p. 5, 1875.

radiating cylinder of woody tissue; while lines radiating towards the circumference indicate the outer radiating cylinder composing the inner bark, formed of thick-walled utricles, or elongated cells, of a quadrangular form, arranged in wedge-shaped masses, divided by very coarse cellular tissue, oblong in its transverse section, somewhat like that described by me as occurring in *Calamodendron commune*, and containing a vascular bundle, also wedge-shaped, but increasing in the direction opposite to that in which the first-named wedge-shaped masses do: all figured of the natural size. The outer bark had been converted into a mass of bright coal, about an inch in thickness. Fig. 2 shows the outside appearance of the fossil in a decorticated state, marked with fine longitudinal striæ, irregular ribs and furrows, and some rather indistinct traces of the cicatrices of leaf-scars, which would induce many collectors of coal-plants to class it with a decorticated specimen of *Sigillaria*. The outer bark of the specimen remains attached to it, in the form of coal, united to the matrix of the fossil. The reverse side of the specimen has the same characters, with the exception of the oval protuberance shown in the plate.

In Plate XX, fig. 1, is represented a transverse section of the light-coloured disk previously alluded to and shown, of natural size, in Plate XIX, fig. 1, but here magnified $4\frac{1}{2}$ diameters, exhibiting the central axis composed of hexagonal tubes arranged without order, and of several sizes, those in the middle being rather smaller, but becoming larger towards the outside, where they come in contact with the internal radiating cylinder *b*, and then again diminishing in size just at the point of junction. This was no doubt originally cylindrical, like the stem of the plant; but both parts, in the process of petrification, have been altered by pressure to their present forms. It consists of a broad cylinder (*b*), about an inch in diameter, composed of parallel, elongated, tetragonal, or hexagonal tubes, of equal diameter throughout for the greater part of their length, obtuse or rounded at either extremity, and everywhere marked with crowded parallel lines, which are free or anastomosing all over the surface. The tubes towards the axis are of the smallest diameter; they gradually enlarge towards the circumference, where they are the largest, though bundles of small tubes occasionally occur among the larger. This cylinder, which may be called the internal woody system of the plant, is divided into elongated, wedge-shaped masses, pointed at their posterior or inner extremities, and parted by vascular bundles and fine medullary rays, of various breadths, some much narrower than the diameter of the tubes, others considerably broader, but none are conspicuous to the naked eye, except towards the circumference in some few instances. The disarrangement of the tubes of the central axis seems to be the result of the process of mineralization, as similar appearances have not been observed in many other specimens examined, which in that part are in a more perfect state of preservation. The dark and sharp line separating the vessels of the central axis from those of the internal radiating cylinder does not permit us to clearly see the origin of the vascular bundles or medullary rays which undoubtedly traverse the latter.

Fig. 2 represents a longitudinal section through the specimen, extending across the

20

whole of the internal radiating cylinder and the central axis, magnified 12 diameters; (*a*) showing the barred tubes of the central axis; (*b b*) the internal radiating cylinder of barred tubes, at first next the central axis, small, but increasing in size; (*c c*) as they approach the outside. In this section no part of the zone of lax cellular tissue, the outer radiating cylinder, the prosenchymatous tissue forming the inner bark, or of the outer bark, is shown; nor are there any traces of vascular bundles.

Fig. 3 is a tangential section of a portion of the inner radiating cylinder, magnified 16 diameters; (*d*) showing the large bundle of vascular tubes, and (*d'*) the medullary rays consisting of a single row of cells traversing the inner radiating cylinder.

Although all the tangential sections afford evidence of these two kinds of rays, they have not been yet observed in the longitudinal sections; so we cannot be certain whether the larger have a bundle of vascular tissue surrounded by cellular tissue, like those seen in *Stigmaria*, or not; and we have no direct evidence to connect them with the pith or central axis, the latter being separated from the inner radiating cylinder by a sharp and distinct line, and showing no communication with the pith, such as is seen in some *Stigmariæ*, and in Corda's *Diploxylon cycadoideum*; but exactly resembling Witham's *Anabathra pulcherrima* in every particular.

Fig. 4 is a portion of the outer radiating cylinder, composed of small rectangular tubes, or elongated utricles, magnified 12 diameters. This band of prosenchymatous tissue is traversed by wedge-shaped masses of lax cellular tissue, which gradually diminish in size as they approach the outer bark. The tangential section does not exhibit the vascular or foliar bundles in so good a state of preservation as my large specimen in plate xxxiv, vol. clv, of the 'Phil. Trans.,' but it shows that they are exactly of the same character, so far as they have been preserved.

Fig. 5 (No. 40). This is a transverse section of the inner radiating cylinder, enclosing a central axis or pith, of a small *Sigillaria vascularis*, in all respects, except as to size, similar to the large specimen; magnified 9 diameters. The tubes in the centre, both great and small, are barred on their sides. The specimen has an outer zone of lax cellular tissue, passing into the outer radiating cylinder of prosenchymatous tissue, surrounded by an epidermis converted into coal. The figure is given for the purpose of showing that the small specimens of *Sigillaria vascularis*, with piths of barred vessels, first described by me in the 'Quarterly Journal of the Geological Society,' pass gradually into the large specimens described in the 'Philosophical Transactions' and in the present Monograph.

A number of the central axes have been taken out of the internal radiating cylinders of *Sigillaria vascularis* for the purpose of endeavouring to trace the connection of the medullary rays with the central axes or piths; but no evidence was obtained to show where these rays originated. They could only be traced to the dark line separating the pith from the inner radiating cylinder, but not passing through that line.

The inner walls of the outer radiating cylinder, next the central axis or pith, were found to be finely ribbed and furrowed longitudinally; but no trace of openings, either large or small, similar to those seen on the inner walls of the open-wedged *Stigmariæ*, was met with. In my Memoir in the 'Philosophical Transactions,' figures of three piths, taken out of three specimens of *Sigillaria vascularis*, are given. They consist of barred tubes, and are all alike in their outward appearance, being slightly ribbed and furrowed; but they present no casts of oval openings, such as are seen on the inner sides of the wedge-shaped bundles of the inner radiating cylinder of *Stigmaria* of the open-wedged character.

This specimen in all respects resembles the large specimens of *S. vascularis* described by me in the 'Phil. Transactions,' but the outer radiating cylinder is not shown so well in tangential section as No. 2 therein mentioned. Still sufficient evidence is afforded of the wedge-shaped masses of large and lax parenchymatous tissue, enveloping a kidney-shaped bundle of barred tubes, which traverse and divide the wedge-shaped masses of prosenchymatous tubes or utricles in their way to the leaves. They are not seen in the transverse and longitudinal, only in the tangential, sections; and have not been found to anastomose, as described by MM. Renault and Grand' Eury in the specimen of *Sigillaria spinulosa* described by them.

§ 2. The Specimens Nos. 41, 42, and 43, *Stigmaria ficoides*. Plate XXI, figs. 1—7.

Specimen No. 41, fig. 1 (magnified 6 diameters), is a transverse section of a *Stigmaria ficoides*, found by me in the "Bullion Seam" (marked ** in the section of strata of the Lancashire Coal-measures hereinbefore given at page 12 of this Monograph) at Clough Head, near Burnley, in a calcareous nodule. The specimen is oval, and about an inch in diameter; but the figure only represents the inner radiating cylinder and one of the vascular bundles proceeding to the rootlets, the medulla being absent. The wedge-shaped masses of the wood are parted by wide spaces, and their ends are slightly convex, projecting into the space formerly occupied by the medullary tissue; in every respect similar to Goeppert's specimen, and quite different from the close-wedged example from North Staffordshire described by me in the 'Quarterly Journal of the Geol. Soc.,' which had its woody system separated from the medulla, or central axis, by a sharp and distinct line. The tubes are quadrangular, and arranged in radiating series, being smallest near the axis, and gradually increasing as they approach the circumference. On the left-hand side of the figure there is represented one of the bell-shaped cavities that contained the vascular bundles communicating with the rootlets; but the zone of lax parenchymatous tissue, surrounding the inner radiating cylinder, as well as the outer radiating cylinder, formed of prosenchymatous tissue, first described by me in my Memoir in the 'Philosophical Transactions,' are not shown.

Fig. 2 (magnified 6 diameters) is a longitudinal section of the same specimen, showing the space formerly occupied by the medulla, but now only containing a little disarranged tissue (*a*), the smaller tubes next the medulla, barred on all their sides, forming the inner radiating cylinder (*c*), traversed by one of the large vascular bundles (*d*), proceeding from the medulla towards the outside and the bell-shaped cavity containing the vascular bundle (*d*), traversing the zone of lax parenchymatous tissue (*e*), is well shown on the left-hand side of the figure.

Fig. 3 (magnified 8 diameters) is a tangential section of the same specimen, showing the large oval vascular bundle (*d*), and the numerous small medullary rays of single cells in vertical series (*d'*) traversing the woody cylinder.

Fig. 4 (natural size) is a representation of the outside of the specimen, showing cicatrices of rootlets.

Specimen No. 42, fig. 5 (natural size), shows a beautiful pyritized specimen of *Stigmaria ficoides*,[1] the common open-wedged form, from the Lower Coal-measures of Lancashire (the exact locality not known), exhibiting the inner radiating cylinder in a perfect condition, with the inner ends of the wedge-shaped masses of the woody cylinder pierced by elongated oval cavities, in which were the large vascular bundles communicating with the central axis and proceeding to the rootlets.

Fig. 7, specimen No. 43 (magnified 6 diameters), represents the inner portion of another pyritized specimen from the " Stinking Two Row Seam " of coal at Golden Hill, North Staffordshire, in which the parts of the wedge-shaped masses of the woody cylinder next the medulla not only show the large oval orifices described in No. 42, but also traces of the small medullary rays seen in the tangential sections of the inner parts. This specimen is close-wedged, and is a portion of that described by me in the ' Quarterly Journal of the Geological Society,' and contains the large tubes in the central axis which some writers have taken to be the rootlets of other *Stigmariæ* that have invaded the medullary portion of the plant.

The oval openings (*d*) seen on the ends of the wedge-shaped masses of the woody cylinder are all arranged in quincuncial order, exactly the same as the rootlets are on the outside of the root, and doubtless contained the vascular bundles which proceeded from the medulla and communicated with the rootlets.

---

[1] This is part of a specimen which many years ago led me to search all the Lower Coal-measures of Lancashire and Yorkshire for a long time, until I discovered the calcareous nodules in the " Brooksbottom " and " Upper Foot Coals " near Burnley, Oldham, and Halifax, yielding specimens showing structure, about twenty years since.

§ 3. The Specimen No. 44, *Sigillaria vascularis*, Binney. Plate XXII, figs. 1—4, and Plate XXIII, figs. 1—3.

Specimen No. 44, fig. 1 (natural size), is the outside of a calcareous nodule from the "Bullion Seam" of coal at Clough Head, near Burnley, showing the transverse section of a root, in a most beautiful state of preservation. Every part of the central axis (composed of large and small tubes and cells, barred on all their sides, and arranged without order) is entirely preserved in its structure, as is also the woody cylinder surrounding it; about one half of its diameter arranged in radiating series, with no division in its wedges, and separated from the axis by a sharp and distinct line. No oval openings are seen in the thin end of the wedges, like those of the *Stigmaria* (No. 41) hereinbefore described. On the outside of the woody cylinder is a zone of lax parenchymatous tissue, much disarranged, but gradually passing into an outer radiating cylinder of prosenchymatous tissue, which is traversed by numerous bell-shaped cavities, that contain the vascular bundles leading to the rootlets. In fact, we have a *Stigmaria* showing a medulla, or central axis, surrounded by a woody cylinder, but separated from it by a distinct line of demarcation, and having none of the openings communicating with the central axis such as are met with in *Stigmariæ* like No. 41. The scars, in the form of depressed areolæ, on the outside, are not shown, being enveloped in the matrix of carbonate of lime; but the bell-shaped cavities are well exhibited, and sufficient to prove it to be a *Stigmaria*, and in all respects similar in structure to the specimen of *Sigillaria vascularis*, Nos. 39 and 40, hereinbefore described.

Fig. 2 (magnified 6 diameters) is a representation of the central axis and the inner radiating cylinder; the former composed of large and small tubes and cells, barred on all their sides; the latter being found chiefly near the centre, and the outside next to the woody cylinder. The last-named part is composed of rectangular tubes, placed close together, increasing in size as they extend outwards, and in radiating series. This figure is taken by reflected light.

Fig. 3 (magnified 15 diameters) is a transverse section of another portion of the inner radiating cylinder, composed of rectangular tubes, showing a large vascular bundle and a small medullary ray.

Fig. 4 (magnified 12 diameters) is a tangential section of the inner radiating cylinder of barred tubes, showing a large oval vascular bundle of about thirty cells[1] (*d*), and smaller medullary rays of one and two cells each (*d'*), arranged in vertical series.

Plate XXIII, fig. 1 (magnified 8 diameters), is another transverse section of the same specimen, in a different part of the root, and seen by transmitted light, showing the central axis and the inner radiating cylinder, as previously described. It is given for the

---

[1] In the figure only six cells are shown, but more than thirty are seen by a high power.

purpose of showing that different transverse sections of the root give the same result as to the structure of the central axis.

Fig. 2 (magnified 8 diameters) is a transverse section of a portion of the root taken near the outside of the specimen, showing the lax parenchymatous tissue (*e*) passing into the outer radiating cylinder of prosenchymatous tissue, formed of rectangular tubes (*f*), traversed by four bell-shaped cavities (*d*), connected with the pear-shaped bundle of barred tubes, like those found in *Stigmaria* rootlets. In one of these cavities is a rootlet.

Fig. 3 (magnified 8 diameters) is a longitudinal section (unfortunately not very true, being nearly diagonal across the inner radiating cylinder and the central axis) showing the large and small tubes and cells (*a*) which, by the direction of the section, appear like utricles and cells, large and small, barred on all their sides, the small tubes (*b*) next to the central axis and the larger tubes (*e*), also barred on all their sides, forming the inner radiating cylinder. In this figure the small tubes appear more like cells than tubes; but under a high magnifying power they show the bars on their sides, like those seen on the larger tubes.

In all the longitudinal sections of specimens of *Sigillaria vascularis* (Nos. 39 and 40) we have not been able to trace any vascular bundles proceeding from the central axis or traversing the inner radiating cylinder, having their origin in a medulla or a medullary sheath, similar to those found in the *Stigmaria ficoides* with open wedges. In the tangential section we cross the large vascular bundles and small medullary rays, which in their sectional view do not afford any direct evidence of their being formed of barred tubes, and this is all that can be said of them. In the examination of numerous casts of the outside of the medulla, or central axis, of *Sigillaria vascularis*, as previously stated, there have been found no openings in the ends of the wedge-shaped masses of the wood of the inner radiating cylinder, like those found in the open-wedged *Stigmaria*. This is the case with all the specimens of *Diploxylon cycadoideum* and *Sigillaria vascularis* that have come under my observation. In tangential sections of the outer radiating cylinder or inner bark of the latter plant (pl. xxxiv, fig. 2, in my Memoir in the 'Phil. Trans.') there is evidence of a foliar bundle similar to that shown to exist in *Sigillaria spinulosa* by MM. Renault and Grand' Eury. This is also seen in similar sections of the small *Sigillaria vascularis* described by me (*loc. cit.*, pl. xxxv, fig. 5), and which Professor Williamson and Mr. Carruthers think is a *Lepidodendron*, and which Professor Schimper identifies with *L. Veltheimianum*.

The large size of the tubes and cells in the medulla is very remarkable, and in a great measure accounts for the absence of that part of *Stigmaria*; for such bodies were not likely to have been able to resist decomposition for any considerable time; and it also tends to confirm the probability of the large tubes found in the pith of my Staffordshire specimen, hereinbefore referred to and questioned by Professor Williamson.

§ 4. THE SPECIMEN No. 45, *Stigmaria ficoides*. Plate XXIV, figs. 1—3.

Fig. 1 (natural size) represents the exterior of a decorticated *Stigmaria ficoides*, found by me on an old coal-pit hillock at Over Darwen, Lancashire. It is uncertain whether it came from the "Gannister" or from the "Lower Fort Mine" of the Lower Coal-measures, as both those seams had been wrought in the pit. It occurred in a nodule of rich clay-ironstone. The depressed areolæ, with a little mammelon in the centre, marked by a dark spot, as also the corrugated lines surrounding the areolæ, are very distinct, and a better specimen of *Stigmaria ficoides*, of small size, is not often met with, as far as its exterior is concerned.

The rootlet from which my sections were taken was imbedded in the outer radiating cylinder, or inner bark, about half an inch in depth, and was originally one fourth of an inch in diameter, but it had diminished one half, probably from the removal of its thick carbonaceous exterior during the process of petrification. The remaining eighth of an inch is, for the chief part, composed of crystallized matter, most probably silica; and it is only a small circular speck, about one thirtieth of an inch in diameter, in the centre of the rootlet, that affords evidence of structure.

Fig. 2 (magnified 90 diameters) is a transverse section of the small circular speck. Its exterior consists of a ring of fine parenchymatous tissue, three or four cells in breadth. This is surrounded by a space, four or five times the diameter of the ring above named, in which no structure is apparent, the fine tissue formerly occupying it having disappeared; then in the centre there is a beautiful pear-shaped mass of vascular tissue, one ninetieth of an inch in diameter, consisting of twenty-seven large vessels, of hexagonal, pentagonal, and other shapes, and of a bundle of very minute nearly circular vessels at the upper extremity.

Fig. 3 (magnified 90 diameters) is not a straight section, being about half way between a longitudinal and a transverse section; but it clearly proves that the vascular tubes on all their sides were marked with transverse striæ, as described by Professor Goeppert in 1841; but his specimen did not exhibit so many vessels, only eleven, and was not in so good a state of preservation as the one here described. Dr. J. D. Hooker also examined and described a similar specimen, but it was not very distinct. As none of the rootlets thus described were traced to their exact position in the main root; and as in my first description in the 'Quarterly Journal of the Geol. Soc.' no figure was given of the *Stigmaria* in which the rootlet occurred, it has been considered desirable to again describe the specimen and at greater length.

§ 5. The Specimens Nos. 46 and 47, *Stigmaria ficoides.*   Plate XXIV, figs. 4, 5,
No. 46 ; figs. 6, 7, 8, No. 47.

Fig. 4 represents, on a greatly reduced scale, the base of a stem (No. 46) having
the irregular ribs and furrows so generally found on the outside of *Sigillaria vascularis*
when in a decorticated state.   The greatest breadth across the specimen is about
four feet and six inches, from one of the main roots to the other on the opposite side.

Fig. 5, on a greatly reduced scale, represents the under side of the same specimen,
showing the crucial sutures.

Specimen No. 47, fig. 6, on a greatly reduced scale, shows the base of another
decorticated stem, having its sides covered with the same kind of irregular ribs and furrows
as above described on No. 46.   The greatest breadth across the specimen is about thirty
inches.   As part of the secondary roots of this specimen remain, there is evidence of the
same system of dichotomization which has been observed in the Dixon Fold, St. Helens,
and other examples of large fossil trees of *Sigillaria vascularis.*

Fig. 7 represents the under view of the same specimen, showing the crucial
sutures.

Fig. 8 shows the distinct areolæ, with a little round elevation in the centre, and the
convex corrugated lines so commonly found on the outside of *Stigmaria.*

The above specimens are in the Museum of the Leeds Philosophical Society, and were
observed in 1839 by the late Mr. Bowman and myself.   That gentleman made drawings
of them at the time, and the present figures are reduced copies.   The specimens consist
of a fine-grained sandstone ; and we were informed that they had been found in the
Lower Coal-measures near Bradford, Yorkshire.

In alluding to these singular sutures Dr. Hooker, at page 417 of his memoir,[1] says—
" A yet more remarkable and anomalous structure in *Sigillaria* than either that of their
stigmaroid roots or fluted stems was pointed out to me by Mr. Binney.   This is the
curious crucial mark which quarters the base of the trunk.   The *Sigillaria* generally
divides into four main roots at the base, which unite to form the crown of the dome
described by Lindley and Hutton ; and it is along the line of union of these four roots
that these strongly marked lines run, all meeting at the centre of the dome.   I know
nothing analogous to this in recent or fossil botany."

Dr. Schimper observes[2]—" On remarque très-souvent à la face inférieure du tronc
une suture en forme de croix, dont les extrémités correspondent aux angles de bifurcation

---

[1] ' Memoirs of the Geological Survey of Great Britain,' vol. ii, part ii, p. 417.
[2] ' Traité de Paléontologie Végétale,' vol. ii, p. 112.

des quatre racines primaires. Cette suture n'est autre chose qu'une ligne de contact produite par l'epaississement de ces quatre racines (voy. notre pl., fig. 14); elle se continue aussi vers le haut entre ces mêmes racines."

V. Concluding Remarks.

When Brongniart described his *Sigillaria elegans*, the Rev. Mr. Harcourt's *Lepidodendron*, Lindley and Hutton's *Stigmaria*, and Mr. Witham's *Anabathra*, he had before him all the materials then known, for examining the structure of those plants, that the Coal-measures had afforded. Subsequently Corda added the *Diploxylon cycadoideum*. Then Goeppert described his *Stigmaria* with the vascular bundles in the pith. But in all these specimens, except the last, the structure of the piths was more or less wanting. The first time that anything was published as to stems with vascular tubes in their piths was in my paper in the 'Quarterly Journal of the Geological Society,' and this was further extended in my Memoir in the 'Philosophical Transactions,' where were described larger specimens of *Sigillaria vascularis* and *Diploxylon cycadoideum*, all showing structure similar to that of the smaller ones first described, with the exception of the *Diploxylon* having the edges of the woody bundles of the inner radiating cylinder slightly lunette-shaped, and running into the pith, like those described by Corda in his specimen, but in a less degree. Professor King, in his description of Witham's *Anabathra*, shows it to be like *Sigillaria vascularis*, except in the pith, which was not distinctly shown. Each plant had the same medulla, inner radiating cylinder traversed by large and small medullary rays, since termed primary and secondary, the same zone of lax parenchymatous tissue, gradually passing into prosenchyma, and traversed by vascular bundles leading to the leaves, and which, although traced to the outside of the inner radiating cylinder, could not be absolutely proved to be connected with the large medullary rays, and the same outer bark generally converted into bright coal. It was also asserted that the small *Lepidodendroid* stem gradually passed into the irregularly ribbed and furrowed *Sigillaria*, and that the open-wedged *Stigmaria* belonged to *Diploxylon cycadoideum*, as its root, whilst the close-wedged one belonged to *Sigillaria vascularis*. Since these, in this Monograph, Mr. Dawes' specimen of *Lepidodendron Harcourtii* has been described, and shown to contain a medulla of orthosenchymatous tissue, which Mr. Harcourt's specimen did not afford. The question now for consideration is this,—is the latter to be regarded as the type of the structure of *Lepidodendron*, or is the new plant described by me to be so regarded, taking the evidence of internal structure, without regarding the external character. So far as the former goes, it appears to me desirable for the present to limit the genus *Lepidodendron* to the old type; and therefore I object to Mr. Carruthers taking my small specimens as *Lepidodendron*, and Professor Williamson taking my large ones as *Diploxylon vasculare*. My names are only provisional, but I think it better that they should remain until we know more of the fructification of the plant.

In all the large specimens of *Sigillaria vascularis* hitherto observed the zone of lax parenchyma intervening betwixt the inner and outer radiating cylinders is so disturbed that we have been unable to absolutely prove that the vascular bundles which traverse the one are connected with those that traverse the other, however probable it may appear that such is the case. In the small specimen, where this part of the plant is seen in contact with the inner radiating cylinder, and extending to the leaf-scars, it proceeds in a nearly horizontal direction, as previously shown in the woodcut (fig. 5), very differently to the vascular bundle of the *Lepidodendron Harcourtii* (No. 31) described in this Monograph, which at first proceeds from the medullary sheath in a nearly vertical direction, and then makes a gradual curve to the leaf-scar. It appears to me nearly certain, as some authors have suggested, that the large vascular bundles which traverse the inner radiating cylinder, and proceed through the outer one to the leaves, are really foliar bundles, and not medullary rays, and that we must limit the term "medullary ray" to the single- and double-celled rays found in the tangential sections of the inner radiating cylinder.

In examining the structure of Coal-measure Plants we labour under great difficulties, owing to the fragmentary state of the specimens, and we have to collect evidence gradually and with patience. It has never been my practice to pretend to do much more than to collect the best specimens, and to carefully describe them, in accordance with the advice of that great botanist, the late Dr. ROBERT BROWN, who more than once stated to me that such was the course he should recommend, and which he himself would adopt. To other more experienced botanists is left the task of comparing the ancient with the modern flora.

To those who asserted confidently that *Sigillaria vascularis* had a medulla of parenchyma, and not of barred tubes, specimen No. 39, hereinbefore described, is adduced as evidence in favour of my views and against theirs; and to those who contended that *Stigmaria* had a medulla of parenchyma, and not of barred tubes as alleged by me, the specimen No. 44 is brought forward in support of my view that such root had a medulla of barred tubes and cells. Both these specimens, to my mind, appear to prove that *Sigillaria vascularis* had for its root a *Stigmaria* with a medulla of barred tubes and cells similar to those found in its own stem, whatever kind of *Stigmaria* other *Sigillariæ* had for their roots. But up to this time, with the exception of No. 46, to my knowledge, no specimen has been described and figured in such a perfect state of preservation, as to prove satisfactorily the true nature of the pith of the root. This remains to be done.

For all the numerous species of *Sigillaria*—and their number is very great—little evidence has been obtained to prove the nature of their respective roots, either by similarity of structure or absolute connection.

As to the fructification of *Sigillaria*, it appears to me pretty certain that it will prove to be something like that of *Lepidodendron*, except that the structure of the axis of the cone will show that it was composed of barred tubes and cells similar to those found in *Sigillaria vascularis*, and not of orthosenchymatous tissue, as has been proved to be the case in *Lepidodendron Harcourtii*, the first and only *Lepidodendron* in which the structure of both stem and cone has been well ascertained. The structure of the specimen No. 19, described at page 49 of this Monograph, as well as the cone mentioned and described in my paper in the 'Philosophical Transactions,' seem to indicate this as most probable.

The outside of the last-named cone, which came from the nodules in the *roof* and not in the *seam* of the "Upper Foot Coal" near Oldham, where most of the specimens are found, was not so well illustrated in the woodcut as it might have been. The bracts supporting the sporangia are arranged around the column of the cone in vertical series and quincuncial order, differently from those in *Lepidodendron Harcourtii*, and exactly resembling the arrangement of the leaf-scars in *Sigillaria organum*. In fact, the ribs, furrows, and scars shown on the outside of the column of this cone are in all respects similar to those found on a small stem of *S. organum*.

PLATE XIX.

Sigillaria vascularis, Binney.

Fig. 1 (No. 39). Transverse section of a stem from the " Hard Bed " of Coal at North Owram, Yorkshire, showing the pith, internal radiating cylinder, lax cellular tissue, external radiating cylinder, and outer bark. The wedge-shaped spaces of a light colour indicate the lax cellular tissue enveloping the foliar bundles communicating with the leaves. Natural size.

Fig. 2. External view of the same specimen, partly decorticated, showing the irregular ribs and furrows. Natural size.

Plate XIX.

N.º 39.

Fig 1.

Fig 2.

J.N.Fitch del. et lith

R.Fitch imp

PLATE XX.

Sigillaria vascularis, Binney.

Fig. 1 (No. 39). Transverse section of the pith and internal radiating cylinder. Magnified 4½ diameters.

Fig. 2. Longitudinal section of the pith, internal radiating cylinder, and a portion of the outside. Magnified 12 diameters.

Fig. 3. Tangential section of a part of the internal radiating cylinder, showing the vascular bundles and medullary rays. Magnified 16 diameters.

Flg. 4. Longitudinal section of a part of the outside radiating cylinder of elongated cells or utricles. Magnified 12 diameters.

Fig. 5 (No. 40). Transverse section of a small specimen from the " Bullion Coal," near Burnley, showing the pith and internal radiating cylinder. Magnified 9 diameters.

In this and the following plates the same parts of the specimens figured are indicated by the same letters, as follow :—

a. The middle part, showing the central axis or pith, composed of large and small scalariform tubes, or utricles and cells, but occasionally more or less separated by fine orthosenchymatous tissue.

b. The small scalariform tubes forming the inner portion of the woody cylinder.

c. The large scalariform tubes forming the outer portion of the woody cylinder.

d. The vascular bundles proceeding from the inner radiating cylinder, traversing the outer radiating cylinder, and extending to the leaves and rootlets.

d'. The medullary rays, formed of a single or double row of cells, seen in a tangential section of the woody cylinder.

e. The mass of parenchymatous tissue, at first (near the woody axis) of a coarse and lax character, but becoming finer and denser as it proceeds outwards, until it becomes prosenchymatous.

f. The elongated tubes, or utricles, arranged in radiating series, forming the outer zone next the epidermis.

g. The epidermis of the plant, nearly always converted into coal.

Plate XX

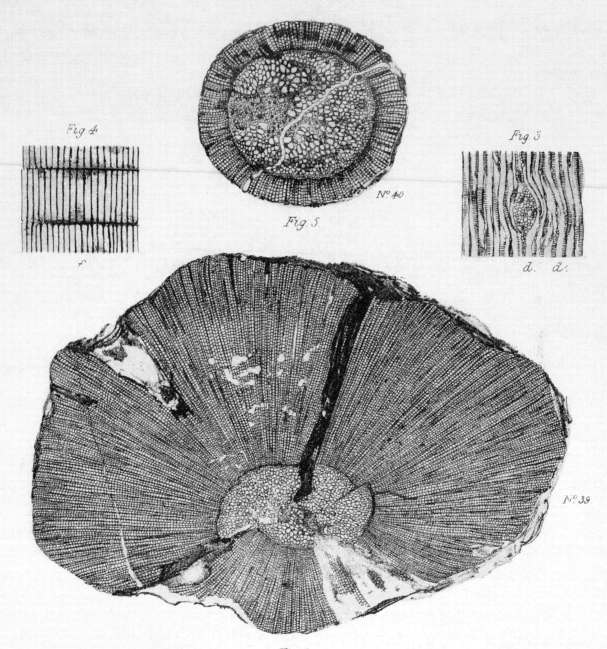

Fig. 4

Fig. 3

N.° 40

Fig. 5

Fig. 1

N.° 39

d.   d.

f.

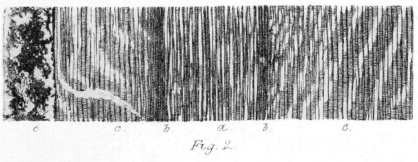

e.   c.   b.   a.   b.   c.

Fig. 2.

J. N. Fitch, del. et lith.   R. Fitch, imp.

PLATE XXI.

Stigmaria ficoides, Lindley and Hutton.

Fig. 1 (No. 41).   A transverse section of a specimen from the " Bullion Seam " of Coal, near Burnley.   Magnified 6 diameters.

Fig. 2.   Longitudinal section of the outer radiating cylinder, and a portion of the outside of the specimen, showing the bell-shaped cavities, from which the rootlets proceed. Magnified 6 diameters.

Fig. 3.   A tangential section of the same specimen, showing the vascular bundles and the medullary rays.   Magnified 8 diameters.

Fig. 4.   The outside of the specimen.   Natural size.

Fig. 5 (No. 42).   A transverse view of a specimen (locality unknown), showing the wedge-shaped masses of the internal woody cylinder, separated by spaces traversed by the vascular bundles.   Natural size.

Fig. 6.   A longitudinal view of the narrow ends of the wedge-shaped masses of the woody cylinder next the pith, traversed by vascular bundles.   Magnified 4 diameters.

Fig. 7 (No. 43).   A longitudinal section of a specimen from Golden Hill, North Staffordshire, showing the narrow ends of the wedges forming the woody cylinder next the pith, and exposing the spaces traversed by the vascular bundles, as well as smaller single-celled orifices, which appear to have been occupied by medullary rays.   Magnified 6 diameters.

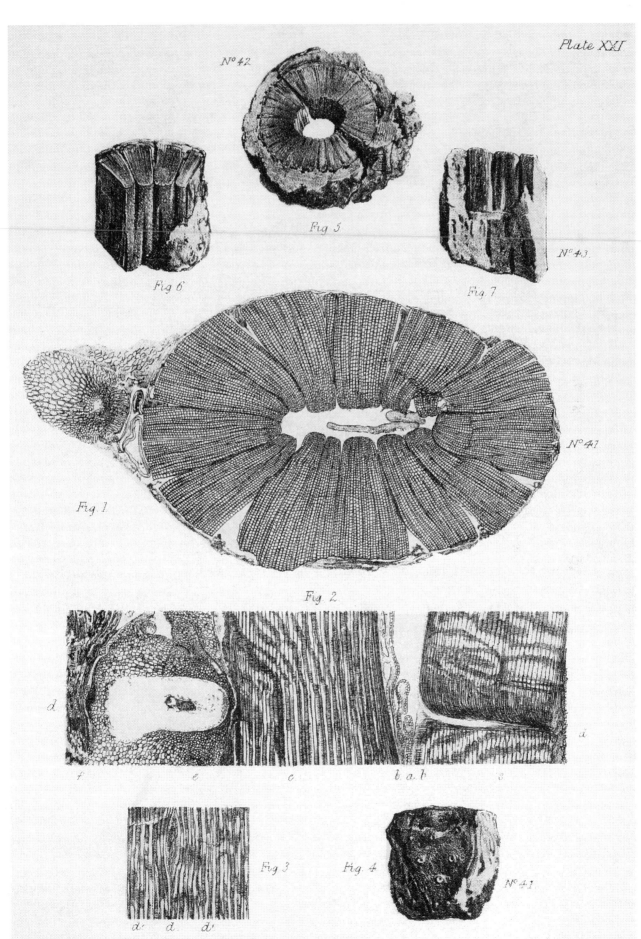

Plate XXI

Nº 42

Fig 5

Fig 6

Fig 7

Nº 43.

Nº 41

Fig 1

Fig. 2

d

f   e   c   b. a. b   3

a

Fig 3   Fig. 4   Nº 44

d'   d   d'

PLATE XXII.

Sigillaria vascularis, Binney.

Fig. 1 (No. 44). A transverse section of a specimen from the " Bullion Seam " of Coal, near Burnley, showing the pith, internal radiating cylinder, cellular tissue, and external radiating cylinder with bell-shaped orifices. Natural size.

Fig. 2. Pith and internal radiating cylinder as seen by reflected light. Magnified 6 diameters.

Fig. 3. Transverse section of a part of the outside of the internal radiating cylinder, showing vascular bundles and spaces where medullary rays may have passed. Magnified 15 diameters.

Fig. 4. Tangential section of the internal radiating cylinder, showing a vascular bundle and medullary rays. Magnified 12 diameters.

Plate XXII

No 44.

Fig 1.

Fig 2.

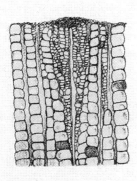

Fig 3

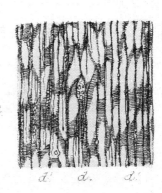

Fig 4.

d¹   d.   d¹

J.N.Fitch. del et lith.

T.R.Shramp

PLATE XXIII.

Sigillaria vascularis, Binney.

Fig. 1. Another transverse section of No. 44, from a different part of the specimen, showing the pith and internal radiating cylinder. Seen by transmitted light. Magnified 8 diameters.

Fig. 2. Transverse section of a portion of the external radiating cylinder, composed of elongated cells (prosenchyma) and parenchymatous tissue, traversed by four bell-shaped orifices, by which the vascular bundles communicated with the rootlets. Magnified 8 diameters.

Fig. 3. Longitudinal section of the pith of small barred cells containing large barred tubes or utricles, and the internal radiating cylinder of barred tubes. Magnified 8 diameters.

Plate XXIII.

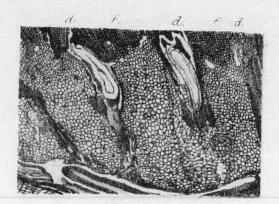

Fig 2.

N° 44.

Fig. 1.

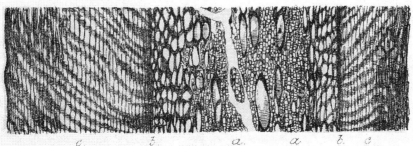

Fig 3.

J.N.Fitch.del.et lith   R.Fitch.imp

PLATE XXIV.

Stigmaria ficoides, Lindl. and Hutton.

Fig. 1 (No. 45). The exterior of a decorticated specimen, from the Lower Coal-measures of Lower Darwen, Lancashire, showing corrugated lines and depressed areolæ, the latter with a circular central elevation, being the outside of the bell-shaped cavities containing the vascular bundles communicating with the rootlets. Natural size.

Fig. 2. Transverse view of the centre of the rootlet, showing the ring of lax cellular tissue, and the vascular bundle in the middle. Magnified 90 diameters.

Fig. 3. Section, partly transverse and partly longitudinal, showing the ring of cellular tissue and the transversely barred vessels of the vascular bundle in the middle. Magnified 90 diameters.

Fig. 4 (No. 46). Side view of a specimen from the Lower Coal-measures, near Bradford, Yorkshire, showing four main roots. Much reduced.

Fig. 5. Under view of the base of the same specimen, showing the crucial sutures. Much reduced.

Fig. 6 (No. 47). Side view of a specimen from the Lower Coal-measures, near Bradford, showing four main roots. Much reduced.

Fig. 7. Under view of the base of the same specimen, showing the crucial sutures. Much reduced.

Fig. 8. A portion of the exterior of one of the main roots, showing the corrugated lines and depressed areolæ, usually found on the outside of *Stigmaria.* Much reduced.

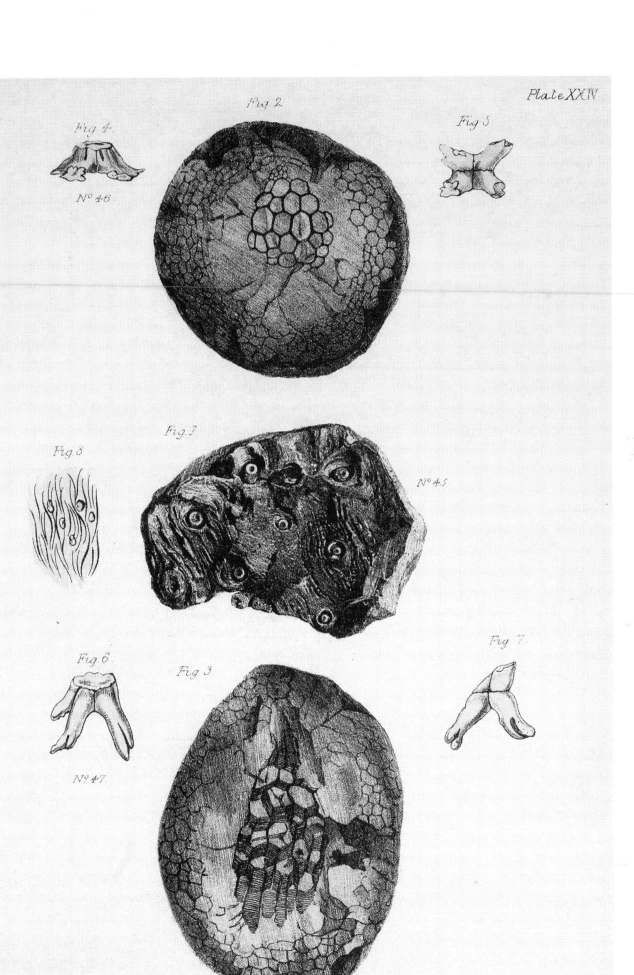

Plate XXIV

Fig 4.

N.º 46

Fig 2

Fig 5

Fig. 1

Fig. 8

N.º 45

Fig. 6

Fig 3

Fig. 7

N.º 47

J. N. Fitch del et lith.

R. Fitch, imp.